Basic Model
Theory

Studies in Logic, Language and Information

The *Studies in Logic, Language and Information* book series is the official book series of the European Association for Logic, Language and Information (FoLLI).

The scope of the book series is the logical and computational foundations of natural, formal, and programming languages, as well as the different forms of human and mechanized inference and information processing. It covers the logical, linguistic, psychological and information-theoretic parts of the cognitive sciences as well as mathematical tools for them. The emphasis is on the theoretical and interdisciplinary aspects of these areas.

The series aims at the rapid dissemination of research monographs, lecture notes and edited volumes at an affordable price.

Basic Model Theory

Kees Doets

CSLI Publications
Center for the Study of Language and Information
Stanford, California
&
FoLLI
The European Association for
Logic, Language and Information

Center for the Study of Language and Information
Leland Stanford Junior University
Printed in the United States
00 99 98 97 96   5 4 3 2 1

Library of Congress Cataloging-in-Publication Data

Doets, Kees.
Basic model theory / Kees Doets
p. cm.
Includes bibliographical references (p. 119–121) and indexes.

ISBN 1-57586-049-X (hardcover : alk. paper). —
ISBN 1-57586-048-1 (pbk : alk. paper)

1. Model theory. I. Title.
QA9.7.D64 1996
511.3—dc20 96-20327
CIP

∞ The acid-free paper used in this book meets the minimum requirements of the American National Standard for Information Sciences—Permanence of Paper for Printed Library Materials, ANSI Z39.48-1984.

Contents

Introduction

This text is written for an audience with some working knowledge of propositional and first-order logic. To make it more self-contained, a natural deduction system and a proof of the completeness theorem are given in Appendix A. Set-theoretic preliminaries are summed up in Appendix B.

The goal of this text is to provide a speedy introduction into what is basic in (mostly: first-order) model theory.

Central results in the main body of this field are theorems like Compactness, Löwenheim-Skolem, Omitting types and Interpolation. From this central area, the following directions sprout:

- model theory for languages extending the first-order ones, abstract model theory,
- applied model theory: non-standard analysis, algebraic model theory, model theory of other special theories,
- recursive model theory,
- finite-model theory,
- classification theory.

There are occasional hints at the first and the fourth, leaving the others largely untouched.

Languages other than first-order discussed below are the following.

- First-order with restricted number of variables,
- (monadic) second-order, admitting quantification over sets of individuals etc.,
- infinitary logic, admitting infinite conjunctions and disjunctions,
- fixed-point logic, which can refer to least fixed points of definable monotone operators.

A short proof of Lindström's famous characterization of first-order logic concludes this introduction.

By then, the ideal student, but hopefully the not-so-ideal student as well, should be comfortable with the standard model theoretic notions introduced here, have some idea concerning the use of Ehrenfeucht's game in simple, concrete situations, and have an impression as to the applicability of some of the basic model theoretic equipment.

Some of the exercises require more of the student than he might be prepared for. Usually, this is indicated by a ♣. Digressions from the main text are printed in a smaller font and separated by horizontal lines.

The bible for the model theory of first-order languages for more than twenty years now is the book *Model Theory* by Chang and Keisler 1990, the last edition of which has been updated. The newer Hodges 1993, that carries the same title, might well rise to the same level of popularity in the near future. These are the books in which you can look for more. In particular, next to Vol. III (Model Theory) of the Ω-*Bibliography of Mathematical Logic* (Ebbinghaus 1987), and the *References for Finite Model Theory* (FMT 1995), these are the places to find bibliographic references, which is the reason that a detailed bibliography is omitted here.

These notes were originally written to accompany a course during the Lisbon 1993 edition of the European Summer School in Logic, Language and Information. (The presence of a course in finite-model theory there accounts for the rather large amount of space devoted to the Ehrenfeucht game in Chapter 3.) Since then, the material has been expanded and used a couple of times for the courses on logic and model theory given at the Mathematics Department of the University of Amsterdam.

Acknowledgments

I thank Dag Westerståhl and an anonymous referee for their valuable criticism of an earlier version of the text, and Maarten de Rijke for his excellent editorial help.

1

Basic Notions

This chapter recalls the basic definitions of first-order model theory. It is assumed that the reader has had some previous contact with these notions. Thus, the purpose of this chapter is mainly to fix the notation and to set the context for the remaining ones.

Let A be a set.

R is an *n-ary relation* over A ($n \geq 1$) if $R \subset A^n$; that is, for all $a_1, \ldots, a_n \in A$ it is in some way determined whether the statement that $R(a_1, \ldots, a_n)$ is *true* or *false*.

f is an *n-ary function over* A ($n \geq 1$) if $f : A^n \to A$; that is, $f(a_1, \ldots, a_n)$ is an element of A whenever $a_1, \ldots, a_n \in A$.

Roughly, a *model* is a complex $\mathcal{A} = (A, R \ldots, f \ldots, a \ldots)$ consisting of a non-empty set A (the *universe* of the model) plus a number of relations $R \ldots$ and functions $f \ldots$ over A and some designated elements (constants) $a \ldots$ from A. A more explicit, vocabulary-related definition is to follow in Definition 1.4.

It is always assumed that A is the universe of $\mathcal{A}$, that B is the one of $\mathcal{B}$, M the one of $\mathcal{M}$, etc.

Examples of models are the familiar structure of the natural numbers $\mathbb{N} = (\mathbb{N}, <, +, \times, 0)$ ($\mathbb{N} = \{0, 1, 2, \ldots\}$; this model has one binary relation $<$, two binary functions $+$ and $\times$, and one constant 0); the structures discussed in algebra: groups, rings, ... etc.

Setting up a system of logic starts by choosing a set of *symbols.* The symbols are partitioned into *logical* and *non-logical* ones. The logical symbols of *first-order* logic are classified into four groups:

1. countably many (individual) *variables*,
2. the equality symbol $=$,
3. symbols for the logical operations: $\neg$, $\wedge$, $\vee$, $\to$, $\leftrightarrow$ (*connectives*) and $\forall$, $\exists$ (*quantifiers*),
4. symbols to indicate grouping: parentheses, comma.

The *non-logical symbols* comprise:

5. relation symbols,
6. function symbols,
7. constant symbols or (individual) constants.

The logical symbols are fixed for every first-order language (although sometimes it is assumed that not all logical operations are present), but the non-logical symbols vary. The set of non-logical symbols is the *vocabulary* of the language.

The different categories of symbols are assumed to be pairwise disjoint, and to every relation and function symbol is associated a positive natural number: the *arity* of the symbol. (You may want to view constant symbols as 0-ary function symbols.)

Given a vocabulary L, you can form *expressions*: finite sequences of L-symbols. Two classes of expressions are singled out: the *L-terms* and the *L-formulas.*

1.1 Terms. All variables and all individual constants of the vocabulary L (considered as length-1 expressions) are L-terms. If $\mathbf{f} \in L$ is an n-ary function symbol and $t_1, \ldots, t_n$ are L-terms, then the sequence $\mathbf{f}(t_1, \ldots, t_n)$ (obtained by writing the terms $t_1, \ldots, t_n$ one after the other, separating them by commas, enclosing the result by parentheses, and putting $\mathbf{f}$ in front) is an L-term.

More precisely, the set of L-terms is the smallest collection containing all variables and constants that is closed under the operation of forming a complex expression $\mathbf{f}(t_1, \ldots, t_n)$ out of expressions $t_1, \ldots, t_n$. So, what you have here is an *inductive definition*, with an accompanying induction principle:

Term Induction. *If X is a set of L-terms that (i) contains all variables and individual constants from L and (ii) contains a term $\mathbf{f}(t_1, \ldots, t_n)$ whenever $\mathbf{f}$ is an n-ary L-function symbol and $t_1, \ldots, t_n \in X$, then X contains all L-terms.*

See Section B.7 for information on inductive definitions in general and Exercise 1 for a first application of term induction.

Next to induct on terms, it is also possible to recursively define operations on them. (The proper justification for this relies on a unique readability result.) An important recursion is the one of Definition 1.6 below. Another one is the notion of a *subterm.* Intuitively, a term s is a *subterm* of the term t if s occurs as a subsequence of consecutive symbols in t.

Subterms. The set $Subt(t)$ of subterms of t is recursively defined by the following clauses.

1. if t is a variable or a constant symbol, then $Subt(t) = \{t\}$,
2. if $t = \mathbf{f}(t_1, \ldots, t_n)$, then $Subt(t) = \{t\} \cup Subt(t_1) \cup \cdots \cup Subt(t_n)$.

1.2 Formulas. (First-order) *L-formulas* are expressions of one of the following forms. 1. $s = t$ (where s and t are L-terms), 2. $\mathbf{r}(t_1, \ldots, t_n)$ (where $\mathbf{r} \in L$ is an n-ary relation symbol and $t_1, \ldots, t_n$ are L-terms), 3. combinations of one of the forms $\neg\varphi$, $(\varphi \wedge \psi)$, $(\varphi \vee \psi)$, $(\varphi \rightarrow \psi)$, $(\varphi \leftrightarrow \psi)$, $\forall x\varphi$, $\exists x\varphi$, where φ and ψ are L-formulas (thought of as formed earlier) and x is a variable.

Formulas of the form $t_1 = t_2$ are called *equalities*; equalities and formulas of the form $\mathbf{r}(t_1, \ldots, t_n)$ are called *atoms*. $\neg\varphi$ is the *negation* of φ, $(\varphi \wedge \psi)$, $(\varphi \vee \psi)$, $(\varphi \rightarrow \psi)$ and $(\varphi \leftrightarrow \psi)$ are the *conjunction*, *disjunction*, *implication* and *equivalence* of φ and ψ; $\forall x\varphi$ and $\exists x\varphi$ are (*universal* and *existential*) *quantifications* of φ with respect to the variable x.

Again, this should be read as an inductive definition, with an accompanying induction principle.

Formula Induction. *Every set of L-formulas that contains the L-atoms and is closed under the logical operations (formation of negations, conjunctions, disjunctions, implications, equivalences and quantifications) contains all formulas.*

Every now and again, variations of this type of induction are used. In every case, such an induction can be viewed as (the "strong form" of) mathematical induction with respect to the number of occurrences of logical constants.

In writing terms and formulas, parentheses (especially, outer ones) will be dropped if this does not lead to confusion. If Φ is a (finite) sequence or set of formulas, then $\bigwedge \Phi$ and $\bigvee \Phi$ can be used to denote the conjunction and disjunction, respectively, of these formulas (formed in any order).

In addition to performing induction on formulas, it is possible to recursively define operations on them. A prominent example of this type of recursion is Definition 1.7. Another one is the definition of the notion of a *subformula* that parallels the one of subterm.

Subformula. The set $Subf(\varphi)$ of *subformulas* of the formula φ is recursively defined by the following clauses.

1. If φ is atomic, then $Subf(\varphi) = \{\varphi\}$,
2. $Subf(\neg\varphi) = \{\neg\varphi\} \cup Subf(\varphi)$, $Subf(\forall x\varphi) = \{\forall x\varphi\} \cup Subf(\varphi)$ (and similarly for existential quantifications),
   $Subf(\varphi \wedge \psi) = \{\varphi \wedge \psi\} \cup Subf(\varphi) \cup Subf(\psi)$ (and similarly for disjunctions, implications and equivalences).

The *scope* of the occurrence of a logical constant in a formula consists of the subformula(s) to which the constant is applied. A quantifier *binds* all occurrences of its variable in its scope — except when such an occurrence is already bound by another quantifier in that scope. For instance, the scope of the quantifier $\forall x$ in the formula $\forall x\,(\mathbf{r}_1(x) \wedge \exists x\,\mathbf{r}_2(x))$ is the subformula $\mathbf{r}_1(x) \wedge \exists x\,\mathbf{r}_2(x)$. It binds the occurrence of x in the subformula $\mathbf{r}_1(x)$. It does not bind the occurrence of x in $\mathbf{r}_2(x)$: this occurrence is bound already by the quantifier $\exists x$.

A variable occurrence that is not bound is *free*.

As to substitution (replacement of (free) occurrences of a variable by a term), substitutability of a term for (the free occurrences of) a variable in a formula (meaning that no variable in the term becomes bound after substitution), see Exercises 3 and 4.

1.3 Sentences. An L-*sentence* is an L-formula in which no variable occurs freely.

The more explicit definition of the notion of a model is the following.

1.4 Models. Let L be a vocabulary. An L-*model* is a pair $\mathcal{A}$ consisting of a non-empty set A, the *universe* of $\mathcal{A}$, and an operation $\sigma \longmapsto \sigma^{\mathcal{A}}$ defined on all non-logical symbols σ of L in such a way that

- if $\mathbf{r} \in L$ is an n-ary relation symbol, then $\mathbf{r}^{\mathcal{A}}$ is an n-ary relation over A,
- if $\mathbf{f} \in L$ is an n-ary function symbol, then $\mathbf{f}^{\mathcal{A}}$ is an n-ary function over A, and
- if $\mathbf{c} \in L$ is a constant symbol, then $\mathbf{c}^{\mathcal{A}} \in A$.

The *cardinality* of a model is the cardinality of its universe. A model is *purely relational* if its vocabulary consists of relation symbols only.

The object $\sigma^{\mathcal{A}}$ is the *interpretation* or *meaning* of σ *in* $\mathcal{A}$, and σ also is called a *name* of $\sigma^{\mathcal{A}}$.

From a certain stage on, symbols and their interpretations shall usually be confused.

Often, an L-model $\mathcal{A}$ over a universe A, where $L = \{\mathbf{r}, \ldots \mathbf{f}, \ldots \mathbf{c}, \ldots\}$, will be represented in the form $\mathcal{A} = (A, \mathbf{r}^{\mathcal{A}}, \ldots \mathbf{f}^{\mathcal{A}}, \ldots \mathbf{c}^{\mathcal{A}}, \ldots)$. And this is the relation with the description of the notion of a model on page 1.

To be able to interpret formulas in a model, you need assignments for their free variables that specify their (temporary) meaning:

1.5 Assignments. Let $\mathcal{A}$ be a model. An $\mathcal{A}$-*assignment* is a function from the set of all variables into the universe A of $\mathcal{A}$.

In the context of a model $\mathcal{A}$, a term stands for an element in $\mathcal{A}$: its *value*, which is calculated (with the help of some assignment) following the way in which the term has been built. (Compare the way polynomials are evaluated in algebra.)

1.6 Value of a Term. Let L be a vocabulary, $\mathcal{A}$ an L-model and α an

$\mathcal{A}$-assignment. For every term t, an element $t^{\mathcal{A}}[\alpha] \in A$, the *value* of t in **A** under α, is defined by the following rules:

1. If t is a variable x: $t^{\mathcal{A}}[\alpha] = \alpha(x)$,
2. if t is a constant symbol $\mathbf{c}$: $t^{\mathcal{A}}[\alpha] = \mathbf{c}^{\mathcal{A}}$,
3. if t has the form $\mathbf{f}(t_1, \ldots, t_n)$, where $\mathbf{f}$ is a function symbol and $t_1, \ldots, t_n$ are terms: $t^{\mathcal{A}}[\alpha] = \mathbf{f}^{\mathcal{A}}(t_1^{\mathcal{A}}[\alpha], \ldots, t_n^{\mathcal{A}}[\alpha])$.

Thus, the value of t is computed by taking a variable to stand for the element given by the assignment and using the meaning of constants and function symbols as supplied by the model.

Next comes the famous "Tarski definition" of the satisfaction relation $\models$ that assigns meanings to formulas: statements about the given model.

1.7 Satisfaction of Formulas. Let L be a vocabulary, $\mathcal{A}$ an L-model and α an $\mathcal{A}$-assignment. For every formula φ, the statement $\mathcal{A} \models \varphi[\alpha]$, *$\varphi$ is satisfied by α in $\mathcal{A}$*, is defined by the following rules:

$$\begin{aligned}
\mathcal{A} \models (s = t)[\alpha] \quad &\Leftrightarrow \quad s^{\mathcal{A}}[\alpha] = t^{\mathcal{A}}[\alpha], \\
\mathcal{A} \models \mathbf{r}(t_1, \ldots, t_n)[\alpha] \quad &\Leftrightarrow \quad \mathbf{r}^{\mathcal{A}}(t_1^{\mathcal{A}}[\alpha], \ldots, t_n^{\mathcal{A}}[\alpha]), \\
\mathcal{A} \models \neg\varphi[\alpha] \quad &\Leftrightarrow \quad \mathcal{A} \not\models \varphi[\alpha], \\
\mathcal{A} \models (\varphi \wedge \psi)[\alpha] \quad &\Leftrightarrow \quad \mathcal{A} \models \varphi[\alpha] \text{ and } \mathcal{A} \models \psi[\alpha] \\
&\quad\quad \text{(similarly for the other connectives)}, \\
\mathcal{A} \models \exists x\varphi[\alpha] \quad &\Leftrightarrow \quad \text{for some } a \in A,\ \mathcal{A} \models \varphi[\alpha^x_a] \\
&\quad\quad \text{(similarly for the } \forall\text{-case)}.
\end{aligned}$$

In the last clause, the notation α^x_a stands for the modification of α that sends the variable x to a (but is otherwise the same as α).

On the use of 1.6 and 1.7. Although these stipulations are called *definitions*, they are of course not as arbitrary as this word may suggest. On the contrary, given the intended meaning of the symbols, they are really unavoidable. However, in concrete, simple situations, you will never need to use them: the value of a simple term always is obvious, as is the meaning of a concrete formula that is not too complicated. The use of 1.6 and 1.7 is in carrying out general arguments that need the principles of term or formula induction. The first spot where such a use is made is in Exercise 2.

1.8 Definitions, Conventions and Notations. If in a given context $x_1, \ldots, x_n$ is a sequence of variables and t a term all of whose variables occur in the sequence, then this can be indicated by writing t as $t(x_1, \ldots, x_n)$. Simultaneously replacing these variables whenever they occur in t by terms $t_1, \ldots, t_n$, the resulting term is then written as $t(t_1, \ldots, t_n)$.

Similarly, a formula φ with free variables among $x_1, \ldots, x_n$ can be written as $\varphi(x_1, \ldots, x_n)$; replacing the *free* occurrences of these variables in φ by $t_1, \ldots, t_n$, the formula obtained will be written as $\varphi(t_1, \ldots, t_n)$.

The result of Exercise 2 permits the following notation. If α is the $\mathcal{A}$-assignment that sends x_i to a_i ($1 \le i \le n$), then the notations $t^{\mathcal{A}}[a_1, \ldots, a_n]$ and $\mathcal{A} \models \varphi[a_1, \ldots, a_n]$ are shorthand for $t^{\mathcal{A}}[\alpha]$ and $\mathcal{A} \models \varphi[\alpha]$, respectively.

The symbol $\models$ is used in a number of different ways.

- The notation $\mathcal{A} \models \varphi$ is used if φ is satisfied in $\mathcal{A}$ by *every* assignment. In that case, we say that φ *is true in* $\mathcal{A}$, that $\mathcal{A}$ *satisfies* φ, or that $\mathcal{A}$ is a *model of* φ. (It is *this* use of the word *model* that is responsible for the term *model theory*.)
- The notation $\models \varphi$ — φ is *logically valid* — is used when φ is true in every model. Formulas are *logically equivalent* if their equivalence is logically valid.
- Finally, if Γ is a set of formulas, the notation $\Gamma \models \varphi$ — φ *follows logically from* Γ — is used in the case that φ is satisfied by an assignment in a model whenever all formulas of Γ are.

From now on, the use of notations such as $t^{\mathcal{A}}[\alpha]$ and $\mathcal{A} \models \varphi[\alpha]$ presupposes that $\mathcal{A}$, t and φ have (or can be assumed to have) the same vocabularies and that α is an $\mathcal{A}$-assignment.

From time to time, logics extending the first-order ones will be considered. Therefore, from now on, the terms *formula* and *sentence* always shall mean *first-order* formula and sentence, respectively, *unless* the contrary is expressly indicated.

Exercises

1 Suppose that t is a term. Let n_i be the number of occurrences of i-ary function symbols in t ($i = 1, 2, 3, \ldots$). Show that the number of occurrences of variables and individual constants in t equals $1 + n_2 + 2n_3 + 3n_4 + \cdots$ ($= 1 + \sum_i (i-1)n_i$).
Hint. Use term induction.

2 The value of a term and the meaning of a formula depend only on the values that are assigned to variables that (freely) occur. More precisely, suppose that $\mathcal{A}$ is a model, α and β are $\mathcal{A}$-assignments, t is a term and φ is a formula. Show that if for all variables x that occur in t and freely occur in φ, $\alpha(x) = \beta(x)$, then

1. $t^{\mathcal{A}}[\alpha] = t^{\mathcal{A}}[\beta]$,
2. $\mathcal{A} \models \varphi[\alpha] \Leftrightarrow \mathcal{A} \models \varphi[\beta]$.

Hint. Apply term and formula induction, respectively.

3 (Substitution and value of a term.) Suppose that s and $t = t(x)$ are terms, $\mathcal{A}$ a model, and α an $\mathcal{A}$-assignment. Let $t(s)$ be the expression obtained from t by replacing all occurrences of x by s.

Show that $t(s)$ is a term.

Next, suppose that $a = s^{\mathcal{A}}[\alpha]$. Show that $t(s)^{\mathcal{A}}[\alpha] = t^{\mathcal{A}}[\alpha^x_a]$.

4 ♣ (Substitution and truth of a formula.) Suppose that s is a term, $\varphi = \varphi(x)$ a formula, $\mathcal{A}$ a model, and α an $\mathcal{A}$-assignment. Let $\varphi(s)$ be the expression obtained from φ by replacing all free occurrences of x by s.

Show that $\varphi(s)$ is a formula.

Next, suppose that $a = s^{\mathcal{A}}[\alpha]$. Show that if s is *substitutable* for x in φ (that is: no free occurrence of x in φ is in the scope of a quantifier that binds a variable from s), then $\mathcal{A} \models \varphi(s)[\alpha] \Leftrightarrow \mathcal{A} \models \varphi[\alpha^x_a]$. Give an example that shows the substitutability condition to be necessary.

5 Suppose that $\mathcal{A}$ is a model in which $a_1, \ldots, a_n \in A$ are the interpretations of the individual constants $\mathbf{c}_1, \ldots, \mathbf{c}_n$ and $\varphi = \varphi(x_1, \ldots, x_n)$ is a formula with $x_1, \ldots, x_n$ free. Show that $\mathcal{A} \models \varphi[a_1, \ldots, a_n]$ iff $\mathcal{A} \models \varphi(\mathbf{c}_1, \ldots, \mathbf{c}_n)$. Thus, in a sense, satisfaction is definable from truth.

6 Suppose that $\varphi = \varphi(x)$ a formula and s a term that is substitutable for x in φ. Let $\varphi(s)$ be the formula obtained from φ by replacing all free occurrences of x by s. Show that

1. $\forall x\varphi \models \varphi(s)$,
2. $\varphi(s) \models \exists x\varphi$.

Hint. Use Exercise 4.

7 Assume that the individual constant $\mathbf{c}$ does occur neither in φ nor in $\psi(x)$. Show the following:

1. if $\varphi \models \psi(\mathbf{c})$, then $\varphi \models \forall x\psi(x)$,
2. if $\psi(\mathbf{c}) \models \varphi$, then $\exists x\psi(x) \models \varphi$.

Hint. Assume that $\varphi \models \psi(\mathbf{c})$ and $\mathcal{A} \models \varphi$. Let $a \in A$ be arbitrary. You have to show, that $\mathcal{A} \models \psi[a]$. Expand $\mathcal{A}$ to a model $(\mathcal{A}, a)$ for a larger vocabulary including $\mathbf{c}$ that interprets $\mathbf{c}$ as a. Of course, it is still true that $(\mathcal{A}, a) \models \varphi$. Since $(\mathcal{A}, a)$ is a model for the vocabulary of $\psi(\mathbf{c})$, it now follows that $(\mathcal{A}, a) \models \psi(\mathbf{c})$. Thus, by Exercise 4, $\mathcal{A} \models \psi[a]$.

8 (An alternative notion of logical consequence.) Sometimes, logical consequence is defined by: $\Gamma \models^* \varphi$ iff φ is true in every model of Γ.

Show that if $\Gamma \models \varphi$, then $\Gamma \models^* \varphi$, and give an example showing that the converse implication can fail. Show that if all elements of Γ are sentences, then $\Gamma \models^* \varphi$ iff $\Gamma \models \varphi$.

1.9 Definability and undefinability of satisfaction. (These explanations are not needed for most of what follows.) In what way do the clauses of 1.7 *define* satisfaction? First of all, they can clearly be satisfied by just one relation $\models$ only. (This is a consequence of formula induction.) This fact can be used to show that, if a structure has a means to code formulas and finite sequences of its elements (which is the case for models of arithmetic and set theory), the satisfaction relation for *first-order* formulas can be *second-order* (see 3.33) defined over it. As they stand, the clauses of 1.7 can be used only to translate a given formula into a statement about the model under consideration. The usual mathematical environment (the "meta-theory") for doing model theory is set theory. (In Chapter 4, this environment comes into play with the subject in a serious way.) Existence of the satisfaction relation in formal set theory should follow from a recursion theorem, and one can wonder as to the precise form of the recursion that is going on. Every recursion employs a well-founded relation. About the well-founded relation this recursion is using there is no doubt: this is the subformula relation. So what should be recursively defined, is, for every formula φ, the set of $\mathcal{A}$-assignments $\|\varphi\| = \|\varphi\|^{\mathcal{A}}$ that satisfy φ. If you view formulas as set-theoretic objects, this recursion takes the following form. Like 1.7, it distinguishes as to the form of the formula. S is the set of $\mathcal{A}$-assignments.

$$\begin{aligned}
\|s = t\| &= \{\alpha \in S \mid s^{\mathcal{A}}[\alpha] = t^{\mathcal{A}}[\alpha]\}, \\
\|\mathbf{r}(t_1, \ldots, t_n)\| &= \{\alpha \in S \mid \mathbf{r}^{\mathcal{A}}(t_1^{\mathcal{A}}[\alpha], \ldots, t_n^{\mathcal{A}}[\alpha])\}, \\
\|\neg\varphi\| &= S - \|\varphi\|, \\
\|\varphi \wedge \psi\| &= \|\varphi\| \cap \|\psi\|, \\
\|\exists x\varphi\| &= \{\alpha \in S \mid \exists a \in A(\alpha_a^x \in \|\varphi\|)\}.
\end{aligned}$$

Such a set-theoretic formalisation of the satisfaction definition allows a comparison with the clauses of 1.7, where these are viewed as a means to translate formulas into statements about the model. What is obtained then is a proof that for every individual formula φ the two ways (translation and definition) of assigning meaning amount to the same thing.

Tarski's adequacy requirement. *For every $\mathcal{A}$-assignment α, $\alpha \in \|\varphi\|^{\mathcal{A}}$ iff $\mathcal{A} \models \varphi[\alpha]$.*

(To explain what this requirement is about, Tarski used the example that the sentence 'snow is white' should be true iff, indeed, snow is white.)

Thus, you can now simply *define* the relation $\models$ that accomplishes the required job by putting $\mathcal{A} \models \varphi[\alpha] :\equiv \alpha \in \|\varphi\|$.

The reason for insisting that the universe of a model be a *set* is now obvious: if A would be a proper class, the values of the operation $\|\cdot\|^{\mathcal{A}}$ would become proper classes as well and the usual set-theoretic instruments would not be sufficient any longer to guarantee its existence.

That is not to say that we never should consider structures over a proper class (the $\in$-structure over the proper class of all sets is the main subject in set theory). In fact, it is not true that you never can define truth in such a structure. For instance, Corollary 3.39 shows that the ordering of the class Ω of all ordinals has a definable notion of truth. However, it is impossible to define truth for the universe of all sets. That (the more general) satisfaction relation of a structure can never be defined over that structure is in fact a very easy application of the *Russell argument* (in the following, read $\sigma(x, y)$ as $y \in x$):

Proposition. *Suppose that $\mathcal{A}$ is an L-structure and that v maps the set of all L-formulas $\varphi = \varphi(x)$ in the one free variable x into A. There is no L-formula $\sigma = \sigma(x, y)$ that defines satisfaction for such formulas, i.e., such that for every*

$\varphi = \varphi(x)$ *and all* $a \in A$*:*

$$\mathcal{A} \models \varphi[a] \ \Leftrightarrow \ \mathcal{A} \models \sigma[v(\varphi), a].$$

Proof. Suppose that $\sigma(x, y)$ satisfies this equivalence. Consider the formula $\varphi := \neg\sigma(x, x)$ and let $a := v(\varphi)$. Then $\mathcal{A} \models \varphi[a]$, according to these definitions, would be tantamount to $\mathcal{A} \models \neg\sigma[v(\varphi), a]$, whereas, according to the equivalence, it should mean that $\mathcal{A} \models \sigma[v(\varphi), a]$. $\dashv$

From this you can deduce the famous Tarski result that truth cannot be defined either, on the assumption that the v-translated notion of substitution is definable. (This assumption is satisfied for the standard structures of arithmetic and set theory and for any reasonable "Gödel numbering" v.)

Note that Gödel's first incompleteness theorem is a corollary of this undefinability result. For instance, it is (tedious but) not particularly difficult to verify that derivability from a (arithmetically) definable system of axioms is (arithmetically) definable and hence differs from (arithmetical) truth. Thus, for any given definable axiom system (Peano arithmetic, Zermelo-Fraenkel set theory, ...) there will be arithmetical truths that are not derivable (unless the system is inconsistent).

2

Relations Between Models

This chapter discusses several basic relations that can exist between two models: isomorphism, (elementary) equivalence and (elementary) embeddability.

2.1 Isomorphism and Equivalence

2.1 Isomorphism. The L-models $\mathcal{A}$ and $\mathcal{B}$ are *isomorphic*, notation: $\mathcal{A} \cong \mathcal{B}$, if there exists an *isomorphism* between them; that is: a bijection $h : A \to B$ between their universes that "preserves structure":

1. for every n-ary relation symbol $\mathbf{r} \in L$ and $a_1, \dots, a_n \in A$:
$$\mathbf{r}^{\mathcal{A}}(a_1, \dots, a_n)) \Leftrightarrow \mathbf{r}^{\mathcal{B}}(h(a_1), \dots, h(a_n)),$$
2. for every n-ary function symbol $\mathbf{f} \in L$ and $a_1, \dots, a_n \in A$:
$$h(\mathbf{f}^{\mathcal{A}}(a_1, \dots, a_n)) = \mathbf{f}^{\mathcal{B}}(h(a_1), \dots, h(a_n)),$$
3. for every individual constant $\mathbf{c} \in L$: $h(\mathbf{c}^{\mathcal{A}}) = \mathbf{c}^{\mathcal{B}}$.

A function $h : A \to B$ is a *homomorphism* from $\mathcal{A}$ to $\mathcal{B}$ if it satisfies conditions 2 and 3 and the $\Rightarrow$-half of condition 1.

An *automorphism* of $\mathcal{A}$ is an isomorphism between $\mathcal{A}$ and $\mathcal{A}$ itself.

Isomorphism is a fundamental mathematical, non-logical concept. The only role of the vocabulary L in the definition is to have a correspondence between relations, functions and constants of the two models. Isomorphic models are *totally indistinguishable* in terms of their structure alone. The next definition introduces first-order logical indistinguishability.

2.2 Equivalence. L-models $\mathcal{A}$ and $\mathcal{B}$ are (*elementarily* or *first-order*) *equivalent*, if they have the same true L-sentences; i.e., if for every L-sentence φ: $\mathcal{A} \models \varphi$ iff $\mathcal{B} \models \varphi$. Notation: $\mathcal{A} \equiv \mathcal{B}$.

Here comes the first theorem.

2.3 Theorem. *If* $\mathcal{A} \cong \mathcal{B}$, *then* $\mathcal{A} \equiv \mathcal{B}$.

Proof. This is the special case of Lemma 2.4.2 where φ is a sentence. ⊣

2.4 Lemma. *If $h : A \to B$ is an isomorphism between $\mathcal{A}$ and $\mathcal{B}$, then, for every term t, formula φ and $\mathcal{A}$-assignment α:*

1. $h(t^{\mathcal{A}}[\alpha]) = t^{\mathcal{B}}[h\alpha]$,
2. $\mathcal{A} \models \varphi[\alpha] \Leftrightarrow \mathcal{B} \models \varphi[h\alpha]$.

Proof. Note that $h\alpha$, the composition of h and α that sends a variable x to $h(\alpha(x))$, is a $\mathcal{B}$-assignment.
1. This is proved by term induction.
If t is a variable x, then (by 1.6.1) $h(x^{\mathcal{A}}[\alpha]) = h(\alpha(x)) = x^{\mathcal{B}}[h\alpha]$.

If t is a constant symbol $\mathbf{c}$, then (by 1.6.2 and 2.1.3) $h(\mathbf{c}^{\mathcal{A}}[\alpha]) = h(\mathbf{c}^{\mathcal{A}}) = \mathbf{c}^{\mathcal{B}} = \mathbf{c}^{\mathcal{B}}[h\alpha]$.

Finally, if $t = \mathbf{f}(t_1, \ldots, t_n)$ and (induction hypothesis) $h(t_i^{\mathcal{A}}[\alpha]) = t_i^{\mathcal{B}}[h\alpha]$ $(i = 1, \ldots, n)$, then

$$\begin{aligned} h(t^{\mathcal{A}}[\alpha]) &= h(\mathbf{f}(t_1, \ldots, t_n)^{\mathcal{A}}[\alpha]) \\ &= h(\mathbf{f}^{\mathcal{A}}(t_1^{\mathcal{A}}[\alpha], \ldots, t_n^{\mathcal{A}}[\alpha])), \text{ (by 1.6.3)} \\ &= \mathbf{f}^{\mathcal{B}}(h(t_1^{\mathcal{A}}[\alpha]), \ldots, h(t_n^{\mathcal{A}}[\alpha])), \text{ (by 2.1.2)} \\ &= \mathbf{f}^{\mathcal{B}}(t_1^{\mathcal{B}}[h\alpha], \ldots, t_n^{\mathcal{B}}[h\alpha])), \text{ (by induction hypothesis)} \\ &= t^{\mathcal{B}}[h\alpha], \text{ (again, by 1.6.3).} \end{aligned}$$

2. This is proved by formula induction.
For an identity $s = t$ we have:

$$\begin{aligned} \mathcal{A} \models (s = t)[\alpha] &\Leftrightarrow s^{\mathcal{A}}[\alpha] = t^{\mathcal{A}}[\alpha], \text{ (by 1.7)} \\ &\Leftrightarrow h(s^{\mathcal{A}}[\alpha]) = h(t^{\mathcal{A}}[\alpha]), \text{ (since } h \text{ is an injection)} \\ &\Leftrightarrow s^{\mathcal{B}}[h\alpha] = t^{\mathcal{B}}[h\alpha], \text{ (by part 1)} \\ &\Leftrightarrow \mathcal{B} \models s = t[h\alpha], \text{ (again, by 1.7).} \end{aligned}$$

For an atom $\mathbf{r}(t_1, \ldots, t_n)$:

$$\begin{aligned} \mathcal{A} \models \mathbf{r}(t_1, \ldots, t_n)[\alpha] &\Leftrightarrow \mathbf{r}^{\mathcal{A}}(t_1^{\mathcal{A}}[\alpha], \ldots, t_n^{\mathcal{A}}[\alpha]), \text{ (by 1.7)} \\ &\Leftrightarrow \mathbf{r}^{\mathcal{B}}(h(t_1^{\mathcal{A}}[\alpha]), \ldots, h(t_n^{\mathcal{A}}[\alpha])), \text{ (by 2.1.1)} \\ &\Leftrightarrow \mathbf{r}^{\mathcal{B}}(t_1^{\mathcal{B}}[h\alpha], \ldots, t_n^{\mathcal{B}}[h\alpha]), \text{ (by part 1)} \\ &\Leftrightarrow \mathcal{B} \models \mathbf{r}(t_1, \ldots, t_n)[h\alpha], \text{ (again, by 1.7).} \end{aligned}$$

For propositional combinations, the induction steps are rather trivial. For instance, assuming the equivalences for φ and ψ by way of induction hypothesis,

$\mathcal{A} \models (\varphi \wedge \psi)[\alpha]$
iff (by 1.7) $\mathcal{A} \models \varphi[\alpha]$ and $\mathcal{A} \models \psi[\alpha]$
iff (by induction hypothesis) $\mathcal{B} \models \varphi[h\alpha]$ and $\mathcal{B} \models \psi[h\alpha]$,
iff (again by 1.7) $\mathcal{B} \models (\varphi \wedge \psi)[h\alpha]$.

Finally, assuming the equivalence for φ as an induction hypothesis (with respect to an *arbitrary* assignment), here follows the equivalence for $\exists x\varphi$:

$\mathcal{A} \models \exists x\varphi[\alpha]$ iff (by Definition 1.7) for some $a \in A$, $\mathcal{A} \models \varphi[\alpha^x_a]$
iff (by induction hypothesis) for some $a \in A$, $\mathcal{B} \models \varphi[h\alpha^x_a]$
iff (since h is a surjection) for some $b \in B$, $\mathcal{B} \models \varphi[h\alpha^x_b]$
iff (again by 1.7) $\mathcal{B} \models \exists x\varphi[h\alpha]$. ⊣

The converse of Theorem 2.3 is *very* false, except for the following.

2.5 Proposition. *If $\mathcal{A} \equiv \mathcal{B}$ and $\mathcal{A}$ is finite, then $\mathcal{A} \cong \mathcal{B}$.*

Proof. By Proposition 3.9 and Theorem 3.18 below. However, you *should* carry out the direct proof indicated in Exercise 10 *now*. ⊣

Exercises

9 Show that an ordering of type $\omega + 1$ is not isomorphic to one of type $\omega + \omega^\star$.
Hint. Use Lemma 2.4.

10 Prove Proposition 2.5.
Hint. Suppose that $\mathcal{A} \equiv \mathcal{B}$, $A = \{a_1, \dots, a_n\}$, but $\mathcal{A} \ncong \mathcal{B}$. Write down a first-order sentence E_n that does not use non-logical symbols with the property that for every model $\mathcal{C}$, $\mathcal{C} \models E_n$ iff C has precisely n elements. Thus, E_n is true of $\mathcal{A}$, true of $\mathcal{B}$, and therefore $\mathcal{B}$ has n elements as well and there are $n!$ bijections between A and B. Show that (since $\mathcal{A} \ncong \mathcal{B}$) for every such bijection $h : A \to B$ there exists a formula $\varphi_h = \varphi_h(x_1, \dots, x_n)$ that is atomic or negated atomic such that $\mathcal{A} \models \varphi_h[a_1, \dots, a_n]$ and $\mathcal{B} \models \neg\varphi_h[h(a_1), \dots, h(a_n)]$. Now the sentence

$$\exists x_1, \dots, x_n \left(\bigwedge_{i<j} x_i \neq x_j \wedge \bigwedge_h \varphi_h \right)$$

is true in $\mathcal{A}$ but false in $\mathcal{B}$, contradicting $\mathcal{A} \equiv \mathcal{B}$.

2.2 (Elementary) Submodels

2.6 Submodel. The L-model $\mathcal{A}$ is a *submodel* of the L-model $\mathcal{B}$, and $\mathcal{B}$ an *extension* or a *supermodel* of $\mathcal{A}$, notation: $\mathcal{A} \subset \mathcal{B}$, if

1. $A \subset B$, and
2. a. $\mathbf{r}^{\mathcal{A}} = \mathbf{r}^{\mathcal{B}} \cap A^n$ whenever $\mathbf{r} \in L$ is an n-ary relation symbol,
   b. $\mathbf{f}^{\mathcal{A}} = \mathbf{f}^{\mathcal{B}} \mid A^n$ (the *restriction* of $\mathbf{f}^{\mathcal{B}}$ to A) whenever $\mathbf{f} \in L$ is an n-ary function symbol,
   c. $\mathbf{c}^{\mathcal{A}} = \mathbf{c}^{\mathcal{B}}$ whenever $\mathbf{c} \in L$ is a constant symbol.

Example. $(\mathbb{N}, +, \times, <, 0, 1) \subset (\mathbb{Q}, +, \times, <, 0, 1) \subset (\mathbb{R}, +, \times, <, 0, 1)$.

From the definition of a submodel it follows that, if $\mathcal{B}$ is a model and $A \subset B$, then A is the universe of a (unique) submodel $\mathcal{A}$ of $\mathcal{B}$ iff A is closed under the functions of $\mathcal{B}$ and contains every constant of $\mathcal{B}$. If A does not satisfy these closure conditions, there always is a smallest set A' such that $A \subset A' \subset B$ that does satisfy them: see Lemma 2.12.

2.7 Lemma *If $\mathcal{A} \subset \mathcal{B}$ and α is an A-assignment, then*

1. *for every term t, $t^{\mathcal{A}}[\alpha] = t^{\mathcal{B}}[\alpha]$,*
2. *for every quantifier-free formula φ, $\mathcal{A} \models \varphi[\alpha] \Leftrightarrow \mathcal{B} \models \varphi[\alpha]$.*

Proof. Exercise 12. $\dashv$

Note that the condition that φ be quantifier-free in 2.7.2 is necessary. For instance, $(\mathbb{N} - \{0\}, <) \subset (\mathbb{N}, <)$; the formula $\exists y(y < x)$ is satisfied by the number 1 in the bigger model (since $0 < 1$), but not in the smaller one. The smaller model is not an *elementary* submodel in the sense of the following definition.

2.8 Elementary Submodel. $\mathcal{A}$ is an *elementary submodel* of $\mathcal{B}$, and $\mathcal{B}$ an *elementary extension* of $\mathcal{A}$; notation: $\mathcal{A} \prec \mathcal{B}$, if $\mathcal{A} \subset \mathcal{B}$, whereas the equivalence 2.7.2 holds for *every* formula φ.

2.9 Remarks.

1. If $\mathcal{A} \prec \mathcal{B}$, then $\mathcal{A} \equiv \mathcal{B}$,
2. if $\mathcal{A}, \mathcal{B} \prec \mathcal{C}$ and $\mathcal{A} \subset \mathcal{B}$, then $\mathcal{A} \prec \mathcal{B}$,
3. if $\mathcal{A} \subset \mathcal{B}$ and $\mathcal{A} \equiv \mathcal{B}$ (or even $\mathcal{A} \cong \mathcal{B}$), then it is not necessary that $\mathcal{A} \prec \mathcal{B}$.

2.10 Tarski's criterion. *If $\mathcal{A} \subset \mathcal{B}$, and for every formula $\varphi = \varphi(x_0, \ldots, x_k)$ and $a_0, \ldots, a_{k-1} \in A$ we have that*

$$\mathcal{B} \models \exists x_k \varphi[a_0, \ldots, a_{k-1}] \Rightarrow \exists a \in A \;\; \mathcal{B} \models \varphi[a_0, \ldots, a_{k-1}, a],$$

then $\mathcal{A}$ is an elementary submodel of $\mathcal{B}$.

Proof. The condition exhibited is the "missing link" in the inductive proof of the required equivalence that for every $\mathcal{A}$-assignment α and every formula φ, $\mathcal{A} \models \varphi[\alpha]$ iff $\mathcal{B} \models \varphi[\alpha]$. Assuming this equivalence for φ as an induction hypothesis, here follows half of the equivalence for $\exists x_k \varphi$.

Suppose that $\mathcal{B} \models \exists x_k \varphi[a_0, \ldots, a_{k-1}]$. By the Tarski condition, for some $a \in A$ we have that $\mathcal{B} \models \varphi[a_0, \ldots, a_{k-1}, a]$. By the induction hypothesis, $\mathcal{A} \models \varphi[a_0, \ldots, a_{k-1}, a]$. Therefore, $\mathcal{A} \models \exists x_k \varphi[a_0, \ldots, a_{k-1}]$.

See Exercise 14 for details. $\dashv$

Notice that the condition of 2.10 refers to satisfaction in the larger model only.

2.11 Example. $(\mathbb{Q},<) \prec (\mathbb{R},<)$. Thus, *order completeness*, or the principle of the least upperbound, cf. 3.35.3 (page 37), (which is a property of the bigger model not shared by the smaller one) is not first-order expressible.

Proof. Exercise 15. $\dashv$

2.12 Lemma. *Suppose that $\mathcal{B}$ is an L-model, that $X \subset B$ and that $\aleph_0, |L|, |X| \leq \mu \leq |B|$. Then a submodel $\mathcal{A} \subset \mathcal{B}$ exists such that $X \subset A$ and $|A| = \mu$.*

Proof. Let $\mathcal{A}$ be the smallest submodel of $\mathcal{B}$ containing some superset of X of power μ. There are several equivalent ways to describe the universe A of such a submodel $\mathcal{A}$. First, choose $A_0 \subset B$ of power μ such that $X \subset A_0$.
1. $A = \bigcup_n A_n$, where $A_0 \subset A_1 \subset A_2 \subset \cdots \subset B$ is the sequence of subsets of B such that A_{n+1} is A_n together with all constants of $\mathcal{B}$ and all values of functions of $\mathcal{B}$ on arguments from A_n.

Note that A contains all constants from $\mathcal{B}$ (indeed, they are elements already of A_1). Furthermore, A is closed under the functions of $\mathcal{B}$. For, let k be the arity of such a function $\mathbf{f}^{\mathcal{B}}$. If $a_1, \ldots, a_k \in A$, then for $n = max\{n_1, \ldots, n_k\}$, where n_i is chosen such that $a_i \in A_{n_i}$ $(i = 1, \ldots, k)$, we have that $a_1, \ldots, a_k \in A_n$, and hence

$$\mathbf{f}^{\mathcal{B}}(a_1, \ldots, a_k) \in A_{n+1} \subset A.$$

Thus, A is the universe of a submodel of $\mathcal{B}$.

Next, note that every A_n has power μ. For A_0 this holds by definition. And if $|A_n| = \mu$, then also $|A_{n+1}| = \mu$, since $|A_{n+1}|$ is the union of the μ-many elements of A_n, the constants of $\mathcal{B}$ of which there are at most μ (since $|L| \leq \mu$) and all values of functions of $\mathcal{B}$ on arguments in A_n — but there are at most μ such functions and at most μ sequences of arguments from A_n. Thus, the power of A equals at most $\mu + \mu + \mu \times \mu = \mu$. (See Section B.5 for these cardinality calculations.)
2. A is the set of values $t^{\mathcal{B}}[\alpha]$ of terms t under assignments α of the variables of t into A_0.
3. A is the *least fixed point* (see Definition B.2 page 113) of the operator that maps a subset Y of B to $Y \cup A_0 \cup \{$all constants from $\mathcal{B}\} \cup \{$all values of functions from $\mathcal{B}$ on arguments from $Y\}$.

See Exercise 18 for further details. $\dashv$

Working slightly harder, we obtain the Downward Löwenheim-Skolem-Tarski Theorem, the first genuine result of first-order model theory.

2.13 Downward Löwenheim-Skolem-Tarski Theorem. *Assume the conditions of 2.12. Then an elementary submodel $\mathcal{A} \prec \mathcal{B}$ exists such that $X \subset A$ and $|A| = \mu$.*

Proof. Pick sets $A_0 \subset A_1 \subset A_2 \subset \cdots \subset B$ of power μ such that $X \subset A_0$ and such that the following condition is satisfied:

for every $\varphi(x_0, \ldots, x_k)$ and $a_0, \ldots, a_{k-1} \in A_n$,
if $\mathcal{B} \models \exists x_k \varphi[a_0, \ldots, a_{k-1}]$, then there exists $a \in A_{n+1}$ such that
$\mathcal{B} \models \varphi[a_0, \ldots, a_{k-1}, a]$.

Put $A := \bigcup_n A_n$. A has power μ, is closed under the functions of $\mathcal{B}$ and contains the constants of $\mathcal{B}$. Thus, A is the universe of a submodel $\mathcal{A}$ of $\mathcal{B}$. In this situation, Tarski's criterion is satisfied.
Here are some details. Assume that $A_0 \subset A_1 \subset \cdots \subset A_n$ of power μ have been found satisfying the requirements. By Exercise 17, over a vocabulary of power $\leq \mu$ there are at most μ formulas, and since there are at most μ finite sequences of elements from A_n, the condition of the construction requires consideration of at most $\mu \times \mu = \mu$ combinations of an existential formula and an assignment for its free variables in A_n. In every combination where the formula happens to be satisfied by the assignment, by the Axiom of Choice pick one satisfying element for the existentially quantified variable. The new set A_{n+1} consists of these elements plus those in A_n. It follows that A_{n+1} has power μ as well. As before, the union $A = \bigcup_n A_n$ also has power μ.

Next, it must be shown that A contains the constants from $\mathcal{B}$ and is closed under the functions from $\mathcal{B}$. Let $\mathbf{c}$ be a constant symbol. Consideration of the formula $\exists x_0(x_0 = \mathbf{c})$ and the empty assignment shows that $\mathbf{c}^{\mathcal{B}}$ already belongs to A_1. Furthermore, let $\mathbf{f}$ be a k-ary function symbol, and let $a_0, \ldots, a_{k-1}$ be a sequence of arguments for the corresponding function $\mathbf{f}^{\mathcal{B}}$ from A. Find n so large that these arguments already belong to A_n. Consideration of the formula $\exists x_k(x_k = \mathbf{f}(x_0, \ldots, x_{k-1}))$ and the assignments $a_0, \ldots, a_{k-1}$ for its free variables shows that $\mathbf{f}^{\mathcal{B}}(a_0, \ldots, a_{k-1})$ is in A_{n+1}.

Finally, Tarski's condition holds. Indeed, if we have that

$$\mathcal{B} \models \exists x_k \varphi[a_0, \ldots, a_{k-1}]$$

(where $a_0, \ldots, a_{k-1} \in A$), then, for some n, $a_0, \ldots, a_{k-1} \in A_n$. Thus, by the constructing condition there exists $a \in A_{n+1}$ such that $\mathcal{B} \models \varphi[a_0, \ldots, a_{k-1}, a]$. $\dashv$

2.14 (Elementary) Embeddings. An *embedding* (*elementary embedding*) of $\mathcal{A}$ into $\mathcal{B}$ is an isomorphism between $\mathcal{A}$ and a submodel (an elementary submodel) of $\mathcal{B}$.

2.15 Lemma. *Assume that h maps the universe of $\mathcal{A}$ into that of $\mathcal{B}$. The following conditions are equivalent:*

1. *h is an embedding (elementary embedding) of $\mathcal{A}$ into $\mathcal{B}$,*
2. *for all atomic — equivalently: for all quantifier-free — (for all) formulas φ and $\mathcal{A}$-assignments α: $\mathcal{A} \models \varphi[\alpha] \Leftrightarrow \mathcal{B} \models \varphi[h\alpha]$.*

Proof. That 1 implies 2 is immediate (for the embedding-case, use Lemmas 2.4 and 2.7). For the other direction, see Exercise 20. $\dashv$

2.16 Chains. Let γ be a limit ordinal. A sequence $(\mathcal{A}_\xi \mid \xi < \gamma)$ of models of length γ is a *chain* (an *elementary chain*) if for all $\xi < \delta < \gamma$ we have that $\mathcal{A}_\xi \subset \mathcal{A}_\delta$ ($\mathcal{A}_\xi \prec \mathcal{A}_\delta$).

$\bigcup_{\xi<\gamma} \mathcal{A}_\xi$, the *limit* of the chain, is the (unique) model with universe $\bigcup_{\xi<\gamma} A_\xi$ that is a supermodel of all models of the chain. (See Exercise 21.)

In the early days of model theory, limits of elementary chains used to be popular. Such constructions are now often replaced by saturation arguments.

2.17 Elementary Chain Lemma. *The limit of an elementary chain elementarily extends all models of the chain.*

Proof. Assume that $\mathcal{A}$ is the limit of the elementary chain of models $(\mathcal{A}_\xi \mid \xi < \gamma)$. Using induction with respect to φ, verify that for all $\xi < \gamma$ and $a_0, \dots, a_{k-1} \in A_\xi$:

$$\mathcal{A}_\xi \models \varphi[a_0, \dots, a_{k-1}] \Leftrightarrow \mathcal{A} \models \varphi[a_0, \dots, a_{k-1}].$$

The only point of interest is the induction step for $\exists$ and $\Leftarrow$: assume that $a_0, \dots, a_{k-1} \in A_\xi$ and $\mathcal{A} \models \exists x_k \varphi[a_0, \dots, a_{k-1}]$. Then $a \in A$ exists such that $\mathcal{A} \models \varphi[a_0, \dots, a_{k-1}, a]$. Say, $a \in A_\delta$. Without loss of generality, it may be assumed that $\delta > \xi$. By induction hypothesis, we have that

$$\mathcal{A}_\delta \models \varphi[a_0, \dots, a_{k-1}, a];$$

and hence, that $\mathcal{A}_\delta \models \exists x_k \varphi[a_0, \dots, a_{k-1}]$. However, $\mathcal{A}_\xi \prec \mathcal{A}_\delta$. It follows that $\mathcal{A}_\xi \models \exists x_k \varphi[a_0, \dots, a_{k-1}]$. $\dashv$

Exercises

11 For every two pairs of models $\mathcal{A}$ and $\mathcal{B}$ from the following list, decide whether (i) $\mathcal{A} \subset \mathcal{B}$, (ii) $\mathcal{A} \cong \mathcal{B}$, (iii) $\mathcal{A}$ is (elementarily) embeddable in $\mathcal{B}$: $(\mathbb{N}, <)$, $(\mathbb{N}, >)$, $(\mathbb{Z}, <)$, $(\mathbb{Z}, >)$, $(\mathbb{N}^+, <)$, $(\mathbb{Z} - \{0\}, <)$, $(\mathbb{Q}, <)$, $(\mathbb{Q} - \{0\}, <)$, $(\mathbb{Q}^+, <)$, $(\mathbb{R}, <)$, $(\mathbb{R} - \{0\}, <)$, $(\mathbb{R}^+, <)$, $(\mathbb{R} - \mathbb{Q}, <)$.
Answer the same question for the models $(\mathcal{A}, 1)$ and $(\mathcal{B}, 1)$ with constant 1, where $\mathcal{A}$ and $\mathcal{B}$ are models from the list.

12 Prove Lemma 2.7.
Hint. Use induction on terms and formulas. In fact, you can take (the appropriate part of) the proof of Lemma 2.4 and just erase h everywhere.

13 Prove the claims from 2.9.

14 Complete the proof of Tarski's criterion 2.10.

15 Prove 2.11.
Hint. Verify Tarski's criterion using the following observation: if r is a real and $q_1, \dots, q_n$ are rational, then an automorphism h of $(\mathbb{R}, <)$ exists such that $h(r)$ is rational and $h(q_i) = q_i$ $(i = 1, \dots, n)$. Apply Lemma 2.4.

16 Show that $(\mathbb{N}^+, 1, 2, 3, \ldots) \prec (\mathbb{N}, 1, 2, 3, \ldots)$. ($\mathbb{N}^+$ is the set of positive natural numbers. The vocabulary of these models has no relation or function symbols and infinitely many constant symbols for the elements of $\mathbb{N}^+$.)
Hint. This is similar to Exercise 15. Note that a formula contains only finitely many constant symbols.

17 Let L be a vocabulary. Show that there are at most $|L| + \aleph_0$ L-terms and L-formulas.

18 Verify the claims from the proof of Lemma 2.12. In particular, why do the three descriptions of the set A all refer to the same thing?

19 Suppose that $\varphi = \varphi(x_0, \ldots, x_k)$. A *Skolem function* for $\exists x_k \varphi$ in $\mathcal{B}$ is a function f over B such that for every $a_0, \ldots, a_{k-1} \in B$, if $\mathcal{B} \models \exists x_k \varphi[a_0, \ldots, a_{k-1}]$, then $\mathcal{B} \models \varphi[a_0, \ldots, a_{k-1}, f(a_0, \ldots, a_{k-1})]$. Using the Axiom of Choice you can construct Skolem functions for every existential formula. Show that, using this, Theorem 2.13 becomes a corollary to Lemma 2.12.

20 Prove the remaining halves of Lemma 2.15.

21 Prove that chains of models do have limits in the sense of Definition 2.16.

Next follow some set-theoretic applications for students familiar with the set-theoretic *cumulative hierarchy*

$$\{V_\alpha\}_{\alpha \in OR}$$

and the *constructible hierarchy*

$$\{L_\alpha\}_{\alpha \in OR}.$$

See Appendix B for explanations. Here, $\mathcal{A} \prec_\Sigma \mathcal{B}$ means that every formula of Σ is satisfied by a given $\mathcal{A}$-assignment in $\mathcal{A}$ iff this is the case in $\mathcal{B}$. If Σ is *closed under subformulas*, i.e., if every subformula of a formula in Σ again belongs to Σ, then the Elementary Chain Lemma 2.17 holds when $\prec$ is replaced by $\prec_\Sigma$.

2.18 The Reflection Principle. *Suppose that Σ is a finite set of set-theoretic formulas, closed under subformulas. ZF proves that the classes of ordinals α for which $(V_\alpha, \in) \prec_\Sigma (V, \in)$ and $(L_\alpha, \in) \prec_\Sigma (L, \in)$, are closed and unbounded.*

22 ♣ Prove the Reflection Principle 2.18. Note that there are but two properties of these hierarchies that are needed for the proof, namely: $\alpha < \beta \Rightarrow V_\alpha \subset V_\beta$, and, for limits γ, $V_\gamma = \bigcup_{\xi<\gamma} V_\xi$ (and similarly for the L_α).

23 Assume that $\alpha < \beta$ are ordinals such that $(V_\alpha, \in) \prec (V_\beta, \in)$. Show that $(V_\alpha, \in) \models ZF$.
Hint. First, show that α is a limit, that $\alpha > \omega$, and that $(V_\alpha, \in)$ is a model of the *Collection Schema* $\forall x \in a\, \exists y\, \varphi \rightarrow \exists b\, \forall x \in a\, \exists y \in b\, \varphi$ (b not free in φ).

24 ♣ Assume that the initial number α has strongly inaccessible cardinality. Show that $\{\beta < \alpha \mid (V_\beta, \in) \prec (V_\alpha, \in)\}$ is closed and unbounded in α.

Show that if $\alpha > \omega$ is uncountable and regular, then $\{\beta < \alpha \mid (L_\beta, \in) \prec (L_\alpha, \in)\}$ is closed and unbounded in α.
Hints. (First part.) Closed: use the Elementary Chain Theorem. Unbounded: if $\beta_0 < \alpha$, define the chain $V_{\beta_0} \subset A_0 \subset V_{\beta_1} \subset A_1 \subset \cdots \subset V_\alpha$ such that $\beta_{n+1} := \bigcap\{\beta \mid A_n \subset V_\beta\}$, $|V_{\beta_n}| = |A_n|$, $(A_n, \in) \prec (V_\alpha, \in)$ (Löwenheim-Skolem-Tarski; note that $\beta_n < \alpha$); now consider $\bigcup_n V_{\beta_n}$.

The following exercise indicates that the relation $\equiv$ between models $(V_\alpha, \in)$ is very much weaker than $\prec$.

25 Show that from the ZF axioms it follows that an unbounded collection C of ordinals exists such that for all $\alpha, \beta \in C$, $(V_\alpha, \in) \equiv (V_\beta, \in)$.
Hint. If you map all ordinals into some *set*, then an unbounded collection of them will be mapped to the same element. Apply this ZF "pigeon-hole principle" to the map that sends an ordinal α to the set of sentences true in $(V_\alpha, \in)$. Thus, the only "logical" ingredient of the argument is the fact that all set-theoretic sentences form a set.

Bibliographic Remarks

The Downward Löwenheim-Skolem Theorem 2.13, as well as the material on elementary submodels, is from Tarski and Vaught 1957. The history of this theorem dates back to Löwenheim 1915 and Skolem 1922.

An old source of results on the "natural" models $(V_\alpha, \in)$ is Montague and Vaught 1959.

3

Ehrenfeucht-Fraïssé Games

The notion of an Ehrenfeucht-Fraïssé game provides a simple characterization of elementary equivalence with straightforward generalizations to several languages other than first-order, which, for simple models (linear orderings, trees, ...), is easy to apply. Besides, it is almost the only technique available in finite-model theory (where Compactness and Löwenheim-Skolem are of no use).

3.1 Finite Games

In order to get neat results, the following is assumed:

3.1 Proviso. *In this section, all vocabularies are finite and do not contain function symbols.*

Warning. In later sections, proofs are often given using the material of this one. So the results there may fall under this proviso as well, even though this restriction on the vocabulary can often be lifted.

For reasons of uniformity, this chapter admits models that have an empty universe.

First a preliminary definition, fixing terminology in the context of shifting vocabularies.

3.2 Expansions. If L and L' are vocabularies such that $L \subset L'$, then L' is an *expansion* of L and L is a *reduct* of L'.

If $L \subset L'$, $\mathcal{A}$ is an L-model, and $\mathcal{B}$ is an L'-model such that $A = B$ and the interpretations of L-symbols in $\mathcal{A}$ and $\mathcal{B}$ coincide, then $\mathcal{B}$ is an L'*-expansion* of $\mathcal{A}$ and $\mathcal{A}$ is the L*-reduct* of $\mathcal{B}$; notation: $\mathcal{A} = \mathcal{B} \mid L$.

If $L' - L$ consists of constant symbols only, the corresponding expansions are called *simple*.

If $\mathcal{A}$ is an L-model and $L' = L \cup \{\mathbf{c}_1, \ldots, \mathbf{c}_n\}$, then the simple L'-expansion of $\mathcal{A}$ that interprets $\mathbf{c}_i$ as $a_i \in A$ $(1 \le i \le n)$ is denoted by $(\mathcal{A}, a_1, \ldots, a_n)$.

See Exercise 5 to see how satisfaction in $\mathcal{A}$ by $a_1, \ldots, a_n \in A$ can be reduced to truth in the simple expansion $(\mathcal{A}, a_1, \ldots, a_n)$, using individual constants to refer to these elements.

Note that if φ is an L-sentence, $L \subset L'$, and $\mathcal{A}$ an L'-model, then φ is an L'-sentence as well and we have that $\mathcal{A} \mid L \models \varphi$ iff $\mathcal{A} \models \varphi$.

3.3 Local Isomorphisms. A *local isomorphism* between models $\mathcal{A}$ and $\mathcal{B}$ of the same vocabulary is a finite relation

$$\{(a_1, b_1), \ldots, (a_n, b_n)\} \subset A \times B$$

such that the simple expansions $(\mathcal{A}, a_1, \ldots, a_n)$ and $(\mathcal{B}, b_1, \ldots, b_n)$ satisfy the same atomic sentences.

Every local isomorphism is a (finite) injection. Often, the models involved are purely relational (i.e., there are no individual constants in the vocabulary). In that case, a local isomorphism is the same as an isomorphism between finite submodels. See Exercise 26.

3.4 Examples.

1. The empty function is a local isomorphism between any two purely relational models.
2. Every restriction of a (local) isomorphism is a local isomorphism.
3. The finite injection $\{(0,0), (2,e), (5,\pi)\}$ is a local isomorphism between $(\mathbb{Z}, <)$ and $(\mathbb{R}, <)$.

The last example illustrates the fact that a local isomorphism does not need to extend to an isomorphism.

3.5 Ehrenfeucht's Game. Let $\mathcal{A}$ and $\mathcal{B}$ be models and $n \in \mathbb{N}$ a natural number. The *Ehrenfeucht game* of length n on $\mathcal{A}$ and $\mathcal{B}$, notation: $E(\mathcal{A}, \mathcal{B}, n)$, consists of the following. There are two players, whom are baptized Di and Sy. (Other names are: I and II, the *Spoiler* and the *Duplicator*.) In a play of the game, the players move alternately. Di is granted the first move; the players are allowed n moves each. A move of Di consists in choosing an element of either A or B. A counter-move of Sy consists in choosing an element of either B (in case Di made her choice in A) or A (in the other case).

At the end of each play of the game, there is one winner. Each play determines n pairs ((*move, countermove*), or vice versa) that make up an n-element relation $h := \{(a_1, b_1), \ldots, (a_n, b_n)\}$ between A and B. By definition, Sy has *won* if h is a local isomorphism between $\mathcal{A}$ and $\mathcal{B}$; and Di has won in case this is false.

Since it is allowed for one or two of the models to be empty, we also stipulate that a player whose turn it is but who cannot move by lack of element, loses.

The idea of the game is best explained by revealing some peculiarities about the characters of the participants that become apparent after playing a couple of example games. Thus, Di sees *differences* all around; each of her moves is accompanied by some exclamation "hey, Sy, look: *here* I've found an extraordinary element in this model you can't find the equal of in the other one!". On the other hand, to Sy every two models appear to be *similar* and every move of Di is countered with some "oh yeah? then what about *this* one!"

In the purely relational case, Sy immediately wins every game $E(\mathcal{A}, \mathcal{B}, 0)$ of length 0, since the empty relation is a local isomorphism. If one of the models is empty and the other one is not, then Di can win any length non-0 game by picking an element from the non-empty model, since this cannot be countered by Sy. However, if both models are empty, Sy wins automatically even if $n > 0$.

3.6 Example. Consider the length-3 game on the models $\zeta := (\mathbb{Z}, <)$ and $\lambda := (\mathbb{R}, <)$. Suppose Di and Sy play as follows:

	Di	Sy	Di	Sy	Di	Sy
$\mathbb{Z}$		2	0			5
$\mathbb{R}$	e			0	π	

The moves make up the finite relation $\{(0,0), (2,e), (5,\pi)\}$, which is the local isomorphism of Example 3.4, and so Sy has won this play.

Of course, the real issue with these games is: whom of the players has a *winning strategy*? For this notion, see Definition B.9, page 116. It follows from Lemma B.10 that for every game $E(\mathcal{A}, \mathcal{B}, n)$, exactly one of the players has a winning strategy. To get some feel for this, try your hand at Exercises 27 and 28. The answers to these exercises can be found in the theory that is to follow, but it is useful to experiment a little first.

3.7 Notation. The situation that Sy has a winning strategy for $E(\mathcal{A}, \mathcal{B}, n)$ is denoted by $\mathsf{Sy}(\mathcal{A}, \mathcal{B}, n)$.

The following results are straightforward.

3.8 Lemma.

1. $\mathsf{Sy}(\mathcal{A}, \mathcal{B}, n) \wedge m \leq n \Rightarrow \mathsf{Sy}(\mathcal{A}, \mathcal{B}, m)$,
2. $\mathsf{Sy}(\mathcal{A}, \mathcal{B}, n) \Rightarrow \mathsf{Sy}(\mathcal{B}, \mathcal{A}, n)$,
3. $\mathcal{A} \cong \mathcal{B} \Rightarrow \forall n\, \mathsf{Sy}(\mathcal{A}, \mathcal{B}, n)$,
4. $\mathsf{Sy}(\mathcal{A}, \mathcal{B}, n) \wedge \mathsf{Sy}(\mathcal{B}, \mathcal{C}, n) \Rightarrow \mathsf{Sy}(\mathcal{A}, \mathcal{C}, n)$.

Proof. Here is the argument for part 4. Assume that σ and τ are winning strategies for Sy in the games of length n on $\mathcal{A}$ and $\mathcal{B}$, respectively, $\mathcal{B}$ and $\mathcal{C}$. In order to win the game on $\mathcal{A}$ and $\mathcal{C}$, Sy does the following. Next to the actual playing of $E(\mathcal{A}, \mathcal{C}, n)$ against Di, he is also busy bookkeeping two plays of $E(\mathcal{A},\mathcal{B},n)$ and $E(\mathcal{B},\mathcal{C},n)$ respectively, in which the Sy-moves are executed by σ and τ, respectively. Suppose that Di starts by playing an element a from $\mathcal{A}$. To this move, Sy applies σ, as though it were a first move in the game $E(\mathcal{A},\mathcal{B},n)$. The answer b produced by σ is given as an input to τ as though it were a first move in $E(\mathcal{B},\mathcal{C},n)$. Finally, the answer c given by τ is returned by Sy as his real answer in the game $E(\mathcal{A},\mathcal{C},n)$. A similar procedure is carried out by Sy when Di moves in $\mathcal{C}$. (In that case, the move is given to τ, τ's answer to σ, and σ's answer is taken as the real answer by Sy in $E(\mathcal{A},\mathcal{C},n)$.) Eventually, the relations built by the winning strategies σ and τ must be local isomorphisms between $\mathcal{A}$ and $\mathcal{B}$, respectively, $\mathcal{B}$ and $\mathcal{C}$. Therefore, their composition will be a local isomorphism as well, and hence Sy wins the play from $E(\mathcal{A},\mathcal{C},n)$.

For the remaining parts, see Exercise 31. ⊣

By Theorem 3.18 the following implies Proposition 2.5.

3.9 Proposition. *Assume that $\mathcal{A}$ has n elements.*

1. *If* $\mathsf{Sy}(\mathcal{A},\mathcal{B},n)$, *then there exists an embedding of $\mathcal{A}$ into $\mathcal{B}$,*
2. *if* $\mathsf{Sy}(\mathcal{A},\mathcal{B},n+1)$, *then* $\mathcal{A}\cong\mathcal{B}$.

Proof. See Exercise 32. ⊣

Below, the game is played on linear orderings. Then, arguments can often be given by induction via the following Splitting Lemma.

This uses the following notation. If $<$ is a linear ordering of A and $a\in A$, then $a\!\uparrow$ denotes the submodel of $(A,<)$ the universe of which is $\{x\in A\mid a<x\}$, and $a\!\downarrow$ denotes the submodel of $(A,<)$ the universe of which is $\{x\in A\mid x<a\}$. (Note that these submodels can be empty.)

3.10 Splitting Lemma. *If $\mathcal{A}$ and $\mathcal{B}$ are linear orderings, then*

$$\mathsf{Sy}(\mathcal{A},\mathcal{B},n+1)$$

iff both

("forth") $\forall a\in A\,\exists b\in B\,[\mathsf{Sy}(a\!\downarrow,b\!\downarrow,n)\wedge\mathsf{Sy}(a\!\uparrow,b\!\uparrow,n)]$, *and*
("back") $\forall b\in B\,\exists a\in A\,[\mathsf{Sy}(a\!\downarrow,b\!\downarrow,n)\wedge\mathsf{Sy}(a\!\uparrow,b\!\uparrow,n)]$.

Proof. First, assume that $\mathsf{Sy}(\mathcal{A},\mathcal{B},n+1)$. Suppose that $a\in A$. Consider a as a first move of Di in the game of length $n+1$. Let $b\in B$ be an answer of Sy given by some winning strategy σ. In the game, n moves remain for both players. Now, σ can be used as a winning strategy in $E(a\!\downarrow,b\!\downarrow,n)$. For, if Di chooses some $x<a$, this can be considered a second move in $E(\mathcal{A},\mathcal{B},n+1)$, and σ returns an answer y that will be necessarily $<b$, etc.

Conversely, suppose that the back-and-forth conditions are satisfied. It then follows, that $\mathsf{Sy}(\mathcal{A},\mathcal{B},n{+}1)$. For, suppose that Di plays $a\in A$. According to "forth" there exists $b\in B$ such that $\mathsf{Sy}(a\!\downarrow,b\!\downarrow,n)$ and $\mathsf{Sy}(a\!\uparrow,b\!\uparrow,n)$. Suppose that σ and τ are winning strategies. Then they combine to one winning strategy for the remaining n-move game $E(\mathcal{A}, \mathcal{B}, n+1)$ that follows the pair of moves (a, b). Indeed, a move in $a\!\downarrow$ or in $b\!\downarrow$ will be answered by σ, whereas τ will take care of moves in $a\!\uparrow$ or $b\!\uparrow$. $\dashv$

For (notations of) orderings and their types, and sums and products, see Section B.3 (page 110).

3.11 Example. *For every n we have that* $\mathsf{Sy}(\lambda, \eta, n)$.

Proof. This is by induction with respect to n, using Lemma 3.10 and the fact that for $\alpha = \lambda$ or $= \eta$, every choice of an element in α splits it as $\alpha = \alpha + \mathbf{1} + \alpha$. Note the peculiarity that the winning strategy does not depend on the length of the game here. $\dashv$

Part 1 of the following lemma should be compared to Exercise 28.1; similarly, for part 2, compare Exercise 28.3. These results will be used in Section 3.3.

3.12 Lemma.

1. $k, m \geq 2^n - 1 \;\Rightarrow\; \mathsf{Sy}(\mathbf{k}, \mathbf{m}, n)$,
2. $m \geq 2^n - 1 \;\Rightarrow\; \mathsf{Sy}(\omega + \omega^\star, \mathbf{m}, n)$..

Proof. 1. We argue by induction with respect to n, using Lemma 3.10.
Basis. $n = 0$.
This case is trivial: the models under consideration are purely relational, therefore, the empty relation is a local isomorphism.
Induction step.
Induction hypothesis: assume the implication holds for n.
Now suppose that $k, m \geq 2^{n+1} - 1$. In order for $\mathsf{Sy}(\mathbf{k}, \mathbf{m}, n+1)$ to hold, it suffices (by Lemma 3.10) to show that for every element i in the universe $\{0, \ldots, k-1\}$ of the linear ordering $\mathbf{k}$ there exists an element j in the universe $\{0, \ldots, m-1\}$ of $\mathbf{m}$ such that $\mathsf{Sy}(i\!\downarrow, j\!\downarrow, n)$ en $\mathsf{Sy}(i\!\uparrow, j\!\uparrow, n)$, and conversely. Therefore, assume that $0 \leq i < k$. Note that $i\!\downarrow\; = \mathbf{i}$ and $i\!\uparrow\; = k - i - 1$. Distinguish three cases as to the location of i: i can be located "in the middle", "at the beginning" or "at the end" of $\mathbf{k}$.

(i) ("In the middle.") $i, k - i - 1 \geq 2^n - 1$.
Claim. There exists j, $0 \leq j < m$ such that $j, m - j - 1 \geq 2^n - 1$.
Proof. $m \geq 2^{n+1}$, and $2^{n+1} - 1 = (2^n - 1) + 1 + (2^n - 1)$. $\dashv$
Take such a j. By induction hypothesis we have that $\mathsf{Sy}(i\!\downarrow, j\!\downarrow, n)$ and $\mathsf{Sy}(i\!\uparrow, j\!\uparrow, n)$.

(ii) ("At the beginning.") $i < 2^n - 1$.
Put $j := i$. Then we have that $\mathsf{Sy}(i\!\downarrow, j\!\downarrow, n)$. Furthermore (since $k, m \geq 2^{n+1} = (2^n - 1) + 1 + (2^n - 1)$ and $i = j < 2^n - 1$) we have that $k - i - 1, m - j - 1 \geq 2^n - 1$, hence it follows from the induction hypothesis that $\mathsf{Sy}(i\!\uparrow, j\!\uparrow, n)$.

(iii) ("At the end.") $k - i - 1 < 2^n$.

Now choose j such that $m - j - 1 = k - i - 1$ and argue as under (ii).

2. See Exercise 33. $\dashv$

3.13 Lemma.

1. $\mathsf{Sy}(\alpha_1, \beta_1, n) \wedge \mathsf{Sy}(\alpha_2, \beta_2, n) \Rightarrow \mathsf{Sy}(\alpha_1 + \alpha_2, \beta_1 + \beta_2, n)$; *more generally:*
2. *If I is a linearly ordered set and, for every $i \in I$, α_i and β_i are orderings such that $\mathsf{Sy}(\alpha_i, \beta_i, n)$, then $\mathsf{Sy}(\sum_{i \in I} \alpha_i, \sum_{i \in I} \beta_i, n)$.*

Proof. See Exercise 34. $\dashv$

3.14 Lemma.

1. *For every n:* $\mathsf{Sy}(\omega, \omega + \zeta, n)$.
2. *For every n:* $\mathsf{Sy}(\zeta, \zeta + \zeta, n)$.

Proof. See Exercise 35. $\dashv$

3.15 Lemma. *Suppose that α is an arbitrary order type ($\alpha \neq 0$ in 3) and $n \in \mathbb{N}$. Then:*

1. $m \geq 2^n - 1 \Rightarrow \mathsf{Sy}(\omega + (\zeta \cdot \alpha) + \omega^\star, \mathbf{m}, n)$ (with $\alpha = 0$, this is 3.12.2),
2. $\mathsf{Sy}(\omega, \omega + \zeta \cdot \alpha, n)$ (with $\alpha = 1$, this is 3.14.1),
3. $\mathsf{Sy}(\zeta, \zeta \cdot \alpha, n)$ (with $\alpha = 2$, this is 3.14.2).

Proof. See Exercise 36. $\dashv$

Exercises

26 Show that every local isomorphism is an injection.

Suppose that $\mathcal{A}$ and $\mathcal{B}$ are purely relational (the vocabulary has only relation symbols). Show that a bijection h between a finite subset of A and a finite subset of B is a local isomorphism between $\mathcal{A}$ and $\mathcal{B}$ iff h is an isomorphism between the submodels of $\mathcal{A}$ and $\mathcal{B}$ that have these subsets as universes.

27 Does either Di or Sy have a winning strategy in the game $E(\zeta, \lambda, 3)$?

(For the notations of the linear orderings involved, see Section B.3.)

28 Whom of the players has a winning strategy in the following games?:

1. $E(\mathbf{6}, \mathbf{7}, 3)$; $E(\mathbf{7}, \mathbf{8}, 3)$,
2. $E(\omega, \zeta, 3)$; $E(\omega, \omega + \omega, 3)$,
3. $E(\mathbf{15}, \omega + \omega^\star, 4)$,
4. $E(\omega, \omega + \zeta, 5)$.

29 The existence of a winning strategy for player Sy can be used to "transfer" truth of a statement for one of the models to the other one. (The explanation for this phenomenon is given in the next section.)

1. Suppose that player Sy has a winning strategy for the game

$$E((A, R), (B, S), 2)$$

and that the relation R is symmetric. Show that S is symmetric as well.

2. Give an example showing that in the above "symmetric" cannot be replaced by "dense".
3. However, the phenomenon *does* hold for "dense" if you assume that Sy has a winning strategy for $E((A, R), (B, S), 3)$.
4. Same questions for "transitive".

Solution for 1. Suppose that $b_1 S b_2$. To show that $b_2 S b_1$ holds as well, let Di and Sy play the game $E((A, R), (B, S), 2)$. Let Sy use his winning strategy, whereas Di plays b_1 and b_2 (without paying attention to the move of Sy). Suppose that the winning strategy of Sy returns the answers a_1 and a_2 from A. Since the strategy is winning, $\{(a_1, b_1), (a_2, b_2)\}$ is a local isomorphism. By the fact that $b_1 S b_2$, we therefore also have that $a_1 R a_2$. However, R is symmetric. Thus, $a_2 R a_1$. But then, $b_2 S b_1$, as well.

30 Suppose that $\mathcal{A}$ is a linear ordering, and that Sy has a winning strategy for the game $E(\mathcal{A}, \mathcal{B}, 3)$. Show that $\mathcal{B}$ also is a linear ordering.

Show that it does not suffice for this to assume that $\mathsf{Sy}(\mathcal{A}, \mathcal{B}, 2)$.

31 Prove the remaining parts of Lemma 3.8.

32 Prove Proposition 3.9.
Hint. In part 1, let Sy use his winning strategy. But how do you ask Di to play?

33 Prove Lemma 3.12.2.
Hint. Use 1. and the fact that any choice of an element in $\alpha := \omega + \omega^\star$ splits it as $\mathbf{k} + \mathbf{1} + \alpha$ or $\alpha + \mathbf{1} + \mathbf{k}$ for some $k \in \mathbb{N}$.

34 Prove Lemma 3.13.

35 Prove Lemma 3.14.
Hint. Use Lemma 3.12 and 3.13.

36 Prove Lemma 3.15.

37 Give a simple necessary and sufficient condition on n and the (possibly infinite) number of elements in the universes of $\mathcal{A}$ and $\mathcal{B}$ under which $\mathsf{Sy}((A, \emptyset), (B, \emptyset), n)$. ($(A, \emptyset)$ is the model with universe A for an empty vocabulary.)

38 ♣ Show that if $|A|$, $|B| \geq 2^n$, then $\mathsf{Sy}((\mathcal{P}(A), \subset), (\mathcal{P}(B), \subset), n)$.

3.2 The Meaning of the Game

The logical meaning of the game is now to be revealed.

3.16 Quantifier Rank. The *quantifier rank* of a formula φ is the maximal number of nested quantifiers in φ, that is: the natural number $qr(\varphi)$ recursively computed as follows.

1. The rank of an atom is 0,
2. $qr(\neg\varphi) = qr(\varphi)$,
3. $qr(\varphi \to \psi) = qr(\varphi \wedge \psi) = \ldots = max(qr(\varphi), qr(\psi))$,
4. $qr(\forall x\varphi) = qr(\exists x\varphi) = qr(\varphi) + 1$.

3.17 n-Equivalence. The models $\mathcal{A}$ and $\mathcal{B}$ are *n-equivalent*, notation: $\mathcal{A} \equiv^n \mathcal{B}$, if they satisfy the same sentences of quantifier rank $\leq n$.

3.18 Main Theorem. $\mathsf{Sy}(\mathcal{A}, \mathcal{B}, n) \Leftrightarrow \mathcal{A} \equiv^n \mathcal{B}$.

Reconsider Exercises 28 and 30 in the light of this result.

The following lemma is an immediate consequence of Definition 3.3.

3.19 Lemma. *Suppose that $\mathcal{A}$ and $\mathcal{B}$ are models, and that $a \in A$, $b \in B$. For a finite injection h such that $Dom(h) \subset A$ and $Ran(h) \subset B$ the following two conditions are equivalent:*

1. *h is a local isomorphism between $(\mathcal{A}, a)$ and $(\mathcal{B}, b)$,*
2. *$h \cup \{(a, b)\}$ is a local isomorphism between $\mathcal{A}$ and $\mathcal{B}$.*

The next lemma describes an obvious inductive condition that characterizes the games for which Sy has a winning strategy. Compare it with Lemma 3.10.

3.20 Lemma. *For $\mathsf{Sy}(\mathcal{A}, \mathcal{B}, n+1)$ it is necessary and sufficient that both*

("forth") $\forall a \in A\, \exists b \in B\, \mathsf{Sy}((\mathcal{A}, a), (\mathcal{B}, b), n)$, and
("back") $\forall b \in B\, \exists a \in A\, \mathsf{Sy}((\mathcal{A}, a), (\mathcal{B}, b), n)$.

Proof. Suppose that $\mathsf{Sy}(\mathcal{A}, \mathcal{B}, n+1)$. Fix a winning strategy σ for Sy in the game $E(\mathcal{A}, \mathcal{B}, n+1)$. Suppose that ("forth") $a \in A$. Consider a as a first move of Di. Now σ produces an answer $b \in B$.
Claim. $\mathsf{Sy}((\mathcal{A}, a), (\mathcal{B}, b), n)$.
Proof. σ can be used as a winning strategy for Sy. To be able to apply σ, Sy pretends to play the game $E(\mathcal{A}, \mathcal{B}, n+1)$ in which already a first pair of moves (a, b) has been played. Going about this way, after the playing of $E((\mathcal{A}, a), (\mathcal{B}, b), n)$ a relation h between A and B has been built. Since σ is winning for Sy in $E(\mathcal{A}, \mathcal{B}, n+1)$, $h \cup \{(a, b)\}$ must be a local isomorphism between $\mathcal{A}$ and $\mathcal{B}$. But then, according to Lemma 3.19, h is a local isomorphism between $(\mathcal{A}, a)$ and $(\mathcal{B}, b)$. Hence, this strategy is winning for Sy.

The converse uses Lemma 3.19 in a similar way; cf. Exercise 39. ⊣

Lemma 3.20 motivates the following definition of a monotone operator Γ over the set of local isomorphisms between two models $\mathcal{A}$ and $\mathcal{B}$.

$(\Gamma)(X) = \{h \mid \forall a \in A\, \exists b \in B\, (h \cup \{(a, b)\} \in X) \wedge \forall b \in B\, \exists a \in A\, (h \cup \{(a, b)\} \in X)\}$.

It will transpire that Sy has a winning strategy in the n-game iff $\emptyset \in \Gamma{\downarrow}n$ (see Exercise 40). By Theorem 3.18 (still to be proved), $\emptyset \in \Gamma{\downarrow}\omega$ iff $\mathcal{A} \equiv \mathcal{B}$.

The least fixed point of this operator is empty. You shall meet its greatest fixed point later on.

Proof of Theorem 3.18. We argue by induction on n.
Basis: $n = 0$. To have quantifier rank ≤ 0 means to be quantifier-free.

$\Leftarrow$: That $\mathcal{A} \equiv^0 \mathcal{B}$ means that $\mathcal{A}$ and $\mathcal{B}$ satisfy the same quantifier-free sentences. Thus, in particular, $\mathcal{A}$ and $\mathcal{B}$ satisfy the same atomic sentences.

$\Rightarrow$: Assume that $\mathsf{Sy}(\mathcal{A}, \mathcal{B}, 0)$. Then $\mathcal{A}$ and $\mathcal{B}$ satisfy the same atomic sentences. Now, use Lemma 3.22 to see that $\mathcal{A}$ and $\mathcal{B}$ satisfy the same quantifier-free sentences.

Induction step. Assume (induction hypothesis) the result for n.

First, suppose that $\mathsf{Sy}(\mathcal{A}, \mathcal{B}, n+1)$. By induction on the number of logical symbols in the quantifier-rank $\leq n+1$ sentence φ, it is shown that $\mathcal{A} \models \varphi \Leftrightarrow \mathcal{B} \models \varphi$. The only case that requires attention is when φ is $\exists x \psi$. So assume that $\mathcal{A} \models \exists x \psi(x)$. For instance, $a \in A$ and $\mathcal{A} \models \psi[a]$; equivalently (see Exercise 5): $(\mathcal{A}, a) \models \psi(\mathbf{c})$, where $\mathbf{c}$ is a new individual constant interpreted by a. Now a can be considered a first move of Di in the length $n+1$ game. Applying Lemma 3.20, you obtain $b \in B$ such that $\mathsf{Sy}((\mathcal{A}, a), (\mathcal{B}, b), n)$. By induction hypothesis, $(\mathcal{A}, a) \equiv^n (\mathcal{B}, b)$. Therefore (note that $\psi(\mathbf{c})$ has quantifier rank $\leq n$), $(\mathcal{B}, b) \models \psi(\mathbf{c})$; i.e., $\mathcal{B} \models \psi[b]$, and it follows that $\mathcal{B} \models \exists x \psi$. (Note that in this argument the induction hypothesis must be applied to simple expansions of the two models.)

Next, assume that $\mathcal{A} \equiv^{n+1} \mathcal{B}$. Suppose that $a \in A$. The intention is to show that $b \in B$ exists such that $(\mathcal{A}, a) \equiv^n (\mathcal{B}, b)$. (By induction hypothesis, it then follows that $\mathsf{Sy}((\mathcal{A}, a), (\mathcal{B}, b), n)$. The same goes for the other way around, so by Lemma 3.20 you obtain that $\mathsf{Sy}(\mathcal{A}, \mathcal{B}, n+1)$, as required.) Suppose that such a b does not exist. Then for every $b \in B$ there exists a quantifier rank $\leq n$ sentence $\varphi_b(\mathbf{c})$ satisfied by $(\mathcal{A}, a)$ but not by $(\mathcal{B}, b)$. Suppose that you can choose these sentences in such a way that $\{\psi_b \mid b \in B\}$ is *finite*. Then the quantifier-rank $\leq n+1$ sentence $\exists x \bigwedge_b \psi_b(x)$ is satisfied in $\mathcal{A}$ but not in $\mathcal{B}$, contradicting hypothesis. That you can manage with finitely many ψ_b follows from Lemma 3.23. $\dashv$

3.21 Remark. Note that the relation $\mathsf{Sy}(\mathcal{A}, \mathcal{B}, n)$ is recursively characterized by the following two equivalences.

1. $\mathsf{Sy}(\mathcal{A}, \mathcal{B}, 0) \Leftrightarrow \emptyset$ is a local isomorphism between $\mathcal{A}$ and $\mathcal{B}$,
2. the equivalence of Lemma 3.20:

$$\mathsf{Sy}(\mathcal{A}, \mathcal{B}, n+1) \Leftrightarrow$$
$$\forall a \in A \exists b \in B \mathsf{Sy}((\mathcal{A}, a), (\mathcal{B}, b), n) \wedge \forall b \in B \exists a \in A \mathsf{Sy}((\mathcal{A}, a), (\mathcal{B}, b), n).$$

Therefore, to prove Theorem 3.18, it suffices to show that the relation $\mathcal{A} \equiv^n \mathcal{B}$ also satisfies these equivalences. See Exercise 41.

To express the relation $\mathcal{A} \equiv^n \mathcal{B}$ as the existence of a winning strategy for Sy in the corresponding game (as proposed by Ehrenfeucht) can be considered as a convenient aid to the imagination: the above characterisation really says all there is to it.

3.22 Lemma, *If $\mathcal{A}$ and $\mathcal{B}$ satisfy the same atomic sentences, then they also satisfy the same quantifier-free sentences.*

Proof. See Exercise 42. $\dashv$

3.23 Lemma. *For every k and n there are only finitely many inequivalent formulas of rank $\leq n$ that have $x_1, \ldots, x_k$ free.*

Proof. The proof is by induction with respect to n (keeping k variable).
Basis, $n = 0$: Recall Proviso 3.1, that our vocabulary has no function symbols and is finite. Thus, there are only finitely many atoms in $x_1, \ldots, x_k$. Every formula of rank 0 is quantifier free, and has an equivalent in disjunctive normal form. Obviously, up to equivalence, there are only finitely many disjunctive normal forms using finitely many atoms.
Induction step, $n+1$: Every rank $\leq n+1$ formula with $x_1, \ldots, x_k$ free has an equivalent disjunctive normal form, the ingredients of which are rank $\leq n$ formulas and formulas $\exists x_{k+1}\varphi$ where φ has rank $\leq n$ and $x_1, \ldots, x_{k+1}$ free. By induction hypothesis, up to equivalence there are only finitely many of those. $\dashv$

You can now see what finiteness of the vocabulary is good for. For instance, let $\mathcal{B}$ be a proper elementary extension of $(\mathbb{N}, 0, 1, 2, \ldots)$. (Every proper extension of this model happens to be an elementary one, but if you do not want to accept this on faith, use Theorem 4.10.) Di can already win the length 1 game on these models by choosing an element of $\mathcal{B}$ outside $\mathbb{N}$. A similar example with $(\mathbb{N}, 0, S)$ (where $S(n) := n+1$) illustrates why you have to exclude function symbols.

There are variations on the Ehrenfeucht-Fraïssé game that are adequate with respect to languages other than first-order. For instance, to get the version for (say: monadic) second-order logic (see Definition 3.33 page 36), Di is allowed to also pick a *subset* of one of the models; Sy then is obliged to counter with a subset from the other one.

A nice variation with applications to intensional logics is the one to formulas with a bounded number of variables. (The relation with intensional logics comes from the fact that standard translations into first-order logic can be carried out with finitely many variables, depending on the type of intensional logic considered.) From the above proof it can be seen that the moves of the players are meant as assignments of elements to variables. Now, modify the game as follows. Let $k \in \mathbb{N}$ be a natural number. Di and Sy are given k *pebbles* each, marked $1, \ldots, k$. A move of Di now consists of placing one of her pebbles on an element of one of the two models; Sy counters by placing his corresponding pebble on an element of the other model. If the length of the game exceeds k, Di runs out of pebbles after her k-th move. She is allowed now to re-use one of her pebbles by simply moving it to some other element (of either model). Sy then counters by re-using his corresponding pebble. When the play is over, the positions of the $2k$ pebbles determine a k-element relation between the models; and Sy again *wins* if this is a local isomorphism. For the k-pebble game, there is the following

3.24 Proposition. Sy *has a winning strategy for the k-pebble game of length n on $\mathcal{A}$ and $\mathcal{B}$ iff $\mathcal{A}$ and $\mathcal{B}$ satisfy the same rank $\leq n$-sentences containing at most k variables.*

In the context of linear orderings, 3 variables suffice.

3.25 Proposition. *If $\mathcal{A}$ and $\mathcal{B}$ are linear orderings with the same valid 3-variable sentences of rank $\leq n$, then $\mathcal{A} \equiv^n \mathcal{B}$.*

Proof. Using induction, it is shown that for every n: if g and h are the locations of at most 3×2 pebbles on $\mathcal{A}$ and $\mathcal{B}$, respectively, such that Sy has a winning strategy in the 3-pebble game of length n at position (g, h), then Sy has a winning strategy in the ordinary game of length n at position (g, h).
Basis: $n = 0$. Trivial.
Induction step. Assume the result for n. Suppose that Sy has a winning strategy in the 3-pebble game of length $n+1$ at position (g, h). Distinguish two cases.

(i) At position (g, h), only 2×2 or less pebbles have been placed. Then each player has at least one free pebble. Thus: for every $a \in A$ there exists $b \in B$ and for every $b \in B$ there exists $a \in A$ such that Sy has a winning strategy in the 3-pebble game of length n at position $(g \cup \{a\}, h \cup \{b\})$. By induction hypothesis: for every $a \in A$ there exists $b \in B$ and for every $b \in B$ there exists $a \in A$ such that Sy has a winning strategy in the ordinary game of length n at position $(g \cup \{a\}, h \cup \{b\})$. But that means that Sy has a winning strategy in the ordinary game of length $n + 1$ at (h, g).

(ii) At position (g, h), all 3×2 pebbles have been used. Suppose that g consists of $a_0 < a_1 < a_2$ and h is $b_0 < b_1 < b_2$. A fortiori, Sy has winning strategies for the two 3-pebble games of length $n + 1$ at the two-pebble positions $((a_0, a_1), (b_0, b_1))$ and $((a_1, a_2), (b_1, b_2))$. The argument under (i) shows that Sy has winning strategies σ and τ, respectively, in the ordinary games of length $n+1$ at positions $((a_0, a_1), (b_0, b_1))$ and $((a_1, a_2), (b_1, b_2))$, respectively. But then, Sy has a winning strategy in the ordinary game of length $n + 1$ at position $((a_0, a_1, a_2), (b_0, b_1, b_2))$ as well: moves $< a_1$ or $< b_1$ are countered using σ, whereas moves $> a_1$ or $> b_1$ are countered using τ. $\dashv$

3.26 Corollary. *On the class of linear orderings, every sentence is equivalent with a three-variable sentence.*

Proof. Via Exercise 104. ⊣

Another modification of the game is obtained by stipulating that Sy wins a play in case the relation built is not a local *isomorphism* but a local *homomorphism*, which is a relation $\{(a_1, b_1), \ldots, (a_n, b_n)\} \subset A \times B$ such that every atomic sentence true in $(\mathcal{A}, a_1, \ldots, a_n)$ is satisfied by $(\mathcal{B}, b_1, \ldots, b_n)$ as well (but not necessarily conversely). Every local homomorphism is a function, but it is not necessarily an injection.

The resulting game E^h relates to *positive* formulas, which are generated from the atomic ones using the logical symbols $\wedge$, $\vee$, $\forall$ and $\exists$ only (thus, $\neg$, $\rightarrow$ and $\leftrightarrow$ are not allowed).

Theorem 3.18 can be now be modified to the following result, the proof of which can be obtained by straightforward adaptation of the former one.

3.27 Theorem. Sy *has a winning strategy for* $E^h(\mathcal{A}, \mathcal{B}, n)$ *iff* $\mathcal{B}$ *satisfies every positive quantifier rank* $\leq n$ *sentence true in* $\mathcal{A}$. ⊣

As a final example of modifying the game, you can mix requirements. Assume that $L' = L \cup \{\mathbf{r}\}$, where $\mathbf{r}$ is some n-ary relation symbol. Stipulate that Sy wins iff the end-product of the play is a local isomorphism with respect to L-structure, and a local homomorphism with respect to $\mathbf{r}$. This determines the $\mathbf{r}$-positive game $E^{\mathbf{r}-pos}$. The game is related to so-called $\mathbf{r}$-*positive* sentences, which only use $\wedge$, $\vee$, $\neg$, $\forall$ and $\exists$ and in which $\mathbf{r}$ occurs in the scope of an *even* number of negation symbols. (The restriction that $\rightarrow$ and $\leftrightarrow$ do not occur is needed to keep the counting of negations straight: $\rightarrow$ and $\leftrightarrow$ contain "hidden" negations.)

Exercises

39 Complete the proof of Lemma 3.20.

40 Show that, for the operator Γ defined by 1 (page 28) the following holds:

$$\{(a_1, b_1), \ldots, (a_k, b_k)\} \in \Gamma{\downarrow}n \text{ iff } \mathsf{Sy}((\mathcal{A}, a_1, \ldots, a_k), (\mathcal{B}, b_1, \ldots, b_k), n).$$

41 Prove Theorem 3.18 using the remark from 3.21.
Hint. The argument may be extracted from the one for Theorem 3.18.

42 Prove Lemma 3.22
Hint. Use induction on quantifier-free sentences.

43 Give a proof of the Splitting Lemma 3.10 using Lemma 3.20 and Lemma 3.13.

44 Give an upper bound for the number inequivalent of quantifier rank $\leq n$ formulas in a vocabulary with k variables in, say, one binary relation symbol and m constant symbols.

The following exercise indicates a proof for Theorem 3.18 that does not change the vocabulary.

45 Suppose that $h = \{(a_1, b_1), \ldots, (a_k, b_k)\} \subset A \times B$ is a local isomorphism. Show that the following are equivalent:

1. if $\varphi = \varphi(x_1, \ldots, x_k)$ has quantifier-rank $\leq n$, then
$$\mathcal{A} \models \varphi[a_1, \ldots, a_k] \Leftrightarrow \mathcal{B} \models \varphi[b_1, \ldots, b_k],$$
2. Sy has a winning strategy in position h in the game of length $k + n$ (with n more moves to go for each player).

46 Prove Proposition 3.24.
Hint. Refine the argument for Theorem 3.18.

47 Formulate and prove a theorem that relates the appropriate version of the Ehrenfeucht game to **r**-*positive* sentences.

48 Modify the Ehrenfeucht game of length n on models $\mathcal{A}$ and $\mathcal{B}$ by requiring that Di always picks her moves from A. Formulate and prove the corresponding modification of Theorem 3.18.

49 Show that every two dense linear orderings without endpoints are elementary equivalent.

3.28 n-local isomorphism. A local isomorphism $h = \{(a_1, b_1), \ldots, (a_k, b_k)\}$ between $\mathcal{A}$ and $\mathcal{B}$ is an *n-local isomorphism* if player Sy has a winning strategy for the remaining n moves in the game $E(\mathcal{A}, \mathcal{B}, k + n)$ in position h.

A local isomorphism is *elementary* if it is an n-local isomorphism for every n.

Hence, every local isomorphism is a 0-local isomorphism. From Theorem 3.18 it follows that a local isomorphism $h = \{(a_1, b_1), \ldots, (a_k, b_k)\}$ is an n-local isomorphism iff for every quantifier rank $\leq n$ formula $\varphi = \varphi(x_1, \ldots, x_k)$: $\mathcal{A} \models \varphi[a_1, \ldots, a_k]$ iff $\mathcal{B} \models \varphi[b_1, \ldots, b_k]$.

A finite part of an isomorphism is an elementary local isomorphism.

50 Which local isomorphisms between $\lambda = (\mathbb{R}, <)$ and $\eta = (\mathbb{Q}, <)$ are elementary? What about arbitrary dense linear orderings without endpoints?

51 Which local isomorphisms between ω and $\omega + \zeta$ are elementary?

52 See Lemma 3.12.2 (page 25).

1. For every $n \geq 1$, produce a sentence φ_n of quantifier rank n that holds in a linear ordering $\mathcal{A}$ iff $\mathcal{A}$ has at least $2^n - 1$ elements.
2. Give a simple condition on m and n that is equivalent with $\mathsf{Sy}(\omega + \omega^\star, \mathbf{m}, n)$.

Solution.
1. If φ is a formula and x a variable that does not occur in φ, then by $\varphi^{<x}$

denote the formula obtained from φ by replacing all quantifiers $\forall y \cdots$ and $\exists y \cdots$ by $\forall y < x \cdots (= \forall y(y < x \rightarrow \cdots))$ and $\exists y < x \cdots (= \exists y(y < x \wedge \cdots))$, respectively. $\varphi^{>x}$ is defined analogously. If x does occur in φ, replace bound variables by others. Check that, for a linear ordering $\mathcal{A}$, $a \in A$, and a sentence φ: $\mathcal{A} \models \varphi^{<x}[a]$ iff $a\downarrow \models \varphi$. (This requires a proof using formula induction of something that is slightly more general.) Now, define: $\varphi_1 = \exists x(x = x)$, $\varphi_{n+1} = \exists x(\varphi_n^{<x} \wedge \varphi_n^{>x})$.

2. $\mathsf{Sy}(\omega + \omega^\star, \mathbf{m}, n) \Leftrightarrow m \geq 2^n - 1$.

Proof. ($\Leftarrow$) Cf. Lemma 3.12.
($\Rightarrow$) Use the fact that $\omega + \omega^\star \models \varphi_n$.

53 See Lemma 3.12.1.

1. For $n \geq 2$ and $k < 2^n - 1$ construct a sentence $\psi_{n,k}$ of quantifier rank $\leq n$ that holds in a linear ordering iff it has exactly k elements.
2. Give a simple condition on m and n that is equivalent to $\mathsf{Sy}(\mathbf{k}, \mathbf{m}, n)$.

Hint. Use Exercise 52. Start with $n = 2$ (then $2^2 - 1 = 3$), and $k = 1$, $k = 2$. Next, suppose that $\psi_{n,k}$ has been defined for $n \geq 2$ en $k < 2^n - 1$. To construct $\psi_{n+1,k}$, distinguish cases $1 \leq k < 2^n - 1$, $k = 2^n - 1$, $2^n - 1 < k < 2^{n+1} - 2$, and $k = 2^{n+1} - 1$.

54 Show that the following two conditions are equivalent:

1. The sentence φ has a logical equivalent of quantifier rank $\leq n$,
2. for every two models $\mathcal{A}$, $\mathcal{B}$ such that $\mathcal{A} \equiv^n \mathcal{B}$: if $\mathcal{A} \models \varphi$, then $\mathcal{B} \models \varphi$.

Conclude that *transitivity* can not be expressed with a sentence of quantifier rank < 3. (Cf. Exercise 29.)

3.3 Applications

3.3.1 Beyond First-order

Here follow some properties that cannot be expressed in first-order terms.

3.29 Definition. A formula $\varphi = \varphi(x)$ in one free variable x *defines* the set $\{a \in A \mid \mathcal{A} \models \varphi[a]\}$ in $\mathcal{A}$; a formula $\varphi = \varphi(x, y)$ in two free variables x, y *defines* the relation $\{(a, b) \in A \times A \mid \mathcal{A} \models \varphi[a, b]\}$ in $\mathcal{A}$.

For instance, the formula $x < y \wedge \neg\exists z(x < z \wedge z < y)$ defines the (successor) relation $n + 1 = m$ in the model $\omega = (\mathbb{N}, <)$.

3.30 Example. *The set of even natural numbers is not definable in* $\omega = (\mathbb{N}, <)$.

Proof. Suppose that φ does define this set. I.e., $\omega \models \varphi[n]$ holds iff n is even. Then the sentence

$$\forall x \forall y(x < y \wedge \neg\exists z(x < z \wedge z < y) \rightarrow (\varphi(x) \leftrightarrow \neg\varphi(z)))$$

holds in ω. Since $\omega \equiv \omega + \zeta$ (cf. Lemma 3.14.1 page 26), this sentence also holds in $\omega + \zeta$. Choose n, m in the ζ-tail of $\omega + \zeta$ such that m is the immediate successor of n. Then we have that $\omega+\zeta \models \varphi[n]$ iff $\omega+\zeta \models \neg\varphi[m]$. Consider the automorphism h of $\omega+\zeta$ for which $h(n) = m$. A contradiction follows using Lemma 2.4. $\dashv$

A set $X \subset \mathbb{N}$ is *co-finite* if $\mathbb{N} - X$ is finite.

For every natural number $n \in \mathbb{N}$ it is possible to write down a formula $\pi_n = \pi_n(x)$ expressing that (an element assigned to) x has exactly n predecessors. Thus, $\omega \models \pi_n[m]$ holds iff $m = n$. It follows that a finite set $A \subset \mathbb{N}$ can be defined in ω by the disjunction $\bigvee_{n\in A} \pi_n$; its complement $\mathbb{N} - A$ is defined by the negation of this formula.

Therefore, all finite and co-finite sets of natural numbers are definable in ω. Conversely:

3.31 Proposition. *Every set definable in $\omega = (\mathbb{N}, <)$ is either finite or co-finite.*

Exercises

55 Prove Proposition 3.31.

3.32 Definition. Let $nb(x, y)$ be the formula

$$(x < y \wedge \neg\exists z(x < z \wedge z < y)) \vee (y < x \wedge \neg\exists z(y < z \wedge z < x)),$$

expressing that x and y are neighbours in the ordering $<$. If S is the relation defined by nb in $\mathcal{A} = (A, <)$, then (A, S) is the *neighbour model* corresponding to $\mathcal{A}$; notation: $\mathcal{A}^{nb}$.

For instance, the relation of ω^{nb} is defined by $|n - m| = 1$.

56 Show the following:

1. $\mathcal{A} \equiv^{n+1} \mathcal{B} \Rightarrow \mathcal{A}^{nb} \equiv^{n} \mathcal{B}^{nb}$,
2. $\mathcal{A} \equiv \mathcal{B} \Rightarrow \mathcal{A}^{nb} \equiv \mathcal{B}^{nb}$.

57 The universe of the model C_m is $\{1, \ldots, m\}$ on which the relation R is defined by $iRj :\equiv |i - j| = 1 \vee (i = 1 \wedge j = m) \vee (i = m \wedge j = 1)$. (Visualize this model by drawing $1, \ldots, m$ on a circle.) Show that if $m \geq 2^n$, then $\mathsf{Sy}(C_m, \zeta^{nb}, n)$.

Hint. After the first two moves, the game reduces to one on successor structures of linear orderings. After a first move $a \in C_m$, "cut" C_m in a (by which a is doubled into elements a' and a''). This operation produces the structure $(\mathbf{m} + \mathbf{1}^{nb}, a', a'')$, in which a' and a'' have become endpoints. Similarly, ζ^{OP}, after a first move $b \in \zeta$, is cut open, producing

$$((\omega + \omega^\star)^{nb}, b', b'').$$

Now, use Lemma 3.12.2 and Exercise 56.

58 Consider the successor relation S on ω defined by $nSm :\equiv n + 1 = m$. Show that the ordering $<$ is not definable in $(\mathbb{N}, S)$.

Hint. S is defined in ω by a modification *suc* of the formula *nb*, i.e.: $(\mathbb{N}, S) = \omega^{suc}$. Now suppose that $LO(x, y)$ would be a definition of the required type. Then the sentences

$$\forall x \forall y (x \neq y \rightarrow LO(x, y) \vee LO(y, x))$$

and

$$\forall x \forall y (LO(x, y) \rightarrow \neg LO(y, x))$$

would hold in $(\mathbb{N}, S)$. *Hence* they would hold in $(\omega + \zeta + \zeta)^{suc}$. Deduce a contradiction.

3.3.2 Second-order

In a *(monadic) second-order language* you can write down everything you could write down before, but now you are also allowed to use *relation* variables $X, Y, Z, \ldots$ occurring in atomic formulas $X(t_1, \ldots, t_n)$ and quantifiers over these variables. The meaning of this new form of expressions in the context of a model $\mathcal{A}$ with a universe A is the following: the new variables stand for relations on A; and the *second-order quantifiers* $\forall X$ and $\exists Y$ are to be read as: "for every relation X (of the appropriate arity) on A", and "for some relation X on A," respectively. Finally, an atom $X(t_1, \ldots, t_n)$ is read as "the tuple $t_1, \ldots, t_n$ is in the relation X".

In a similar way, *function* variables could be quantified over.

More formally, the concept of a second-order formula is most easily obtained by modifying Definition 1.2, page 3, viewing some non-logical symbol as a *variable*:

3.33 Second-order Formulas. The class of *second-order formulas* is obtained by adding the following clause to Definition 1.2 (page 3).

4. If φ is an L-formula and $\sigma \in L$ a relation symbol, then $\forall \sigma \varphi$ and $\exists \sigma \varphi$ are $L - \{\sigma\}$-formulas.

If second-order quantification is permitted only over *unary* relation variables, the resulting formulas are called *monadic* second-order. Such a formula is *universal* if it has the form $\forall \mathbf{r} \varphi$ where $\mathbf{r}$ is a sequence of relation variables and φ is first-order; it is *monadic universal* if these relation variables are unary. Π^1_1 and $\Pi^1_1(mon)$ are the classes of universal, and monadic universal second-order formulas, respectively; their existential counterparts Σ^1_1 and $\Sigma^1_1(mon)$ are defined similarly. (The superscript 1 refers to *second* order — a 0 would refer to *first* order! —, the subscript counts quantifier blocks in the prefix, Π says the first block consists of *universal*, Σ says it consists of *existential* quantifiers.)

Although the semantics of this formalism is pretty obvious, here follows the clause that says how to read a universal second-order quantifier. In this clause, α is an $\mathcal{A}$-assignment for the free first-order variables in φ; in the right-hand side, R is the interpretation of the monadic relation "variable"

r in the expansion $(\mathcal{A}, R)$.

$$\mathcal{A} \models \forall \mathbf{r}\varphi[\alpha] \Leftrightarrow \text{ for all } R \subset A,\ (\mathcal{A}, R) \models \varphi[\alpha].$$

We leave it to the reader to write down the other necessary clauses extending Definition 1.7.

A property (of models, or of elements in a given model) that is both Π^1_1 and Σ^1_1 (respectively, $\Pi^1_1(mon)$ and $\Sigma^1_1(mon)$) definable is said to be Δ^1_1 (respectively, $\Delta^1_1(mon)$).

A class of finite models (suitably coded as sequences of symbols) is in NP if membership in the class is **N**on-deterministically Turing machine decidable in **P**olynomial time. Cf. the discussion in A.12. The following result explains the relationship with second-order definability.

3.34 Theorem. *On the class of finite models:* $\Sigma^1_1 = \mathrm{NP}$. $\dashv$

Below, we deal especially with monadic second-order sentences that are *universal*, i.e., have the form $\forall X \varphi$ where φ uses the unary relation variable X in atoms $X(t)$, but it does not quantify over such second-order variables itself.

Instead of $X(t)$, we often write $t \in X$. Notations such as $\forall x \in X\ \varphi$ and $\exists x \in X\ \varphi$ are abbreviations for $\forall x(x \in X \to \varphi)$ and $\exists x(x \in X \wedge \varphi)$, respectively. Using such sentences you can express a lot of things you cannot express in first-order terms.

3.35 Examples. 1. Compare 3.30. The set of even natural numbers can be defined in ω using a monadic second-order formula. An example of such a second-order definition is $\forall X(\Phi(X) \to x \in X)$, where Φ is the formula

$$\begin{aligned}&\exists y \in X \neg\exists z(z < y)) \wedge \\ &\qquad \forall y \forall z(y < z \wedge \neg\exists u(y < u \wedge u < z) \to (y \in X \leftrightarrow z \notin X)).\end{aligned}$$

("The least element is in X, and from an element and its immediate successor exactly one is in X.") The formula Φ is satisfied by the set of even natural numbers only. As a consequence, $\omega \models \Phi[n]$ holds iff n is even:

($\Rightarrow$) If $\omega \models \forall X(\Phi \to x \in X)[n]$, then in particular we have that $\omega \models (\Phi \to x \in X)[n, E]$, where E is the set of even natural numbers. However, E satisfies the left-hand side Φ of this implication. Therefore, E also satisfies the right-hand side, i.e., $n \in E$, n is even.

($\Leftarrow$) If n is even and $A \subset \mathbb{N}$ is a set satisfying Φ, then obviously $A = E$. Indeed, the first conjunct of Φ says that $0 \in A$, the second conjunct has the effect that $1 \notin A$, next, that $2 \in A$, $3 \notin A$, etc. Therefore, $n \in A$.

2. The orderings ω and $\omega + \zeta$ cannot be distinguished by means of a first-order sentence: they are elementarily equivalent. However, consider the following monadic second-order sentence:

$$\forall X[\forall x(\forall y < x(y \in X) \to x \in X) \to \forall x(x \in X)].$$

This sentence (the *Principle of Strong Induction*) expresses a fundamental property of ω, but is false in $\omega + \zeta$. (Consider the initial of type ω of this model as value for X.)
3. The elementary equivalent orderings $(\mathbb{Q}, \leq)$ and $(\mathbb{R}, \leq)$ are distinguished by the monadic second-order *Principle of the Least Upper Bound* ("every non-empty set of reals that has an upper bound also has a least upper bound"):

$$\forall X[\exists x(x \in X) \wedge \exists y \forall x \in X(x \leq y) \rightarrow \exists y(\forall x \in X(x \leq y) \wedge \forall y'(\forall x \in X(x \leq y') \rightarrow y \leq y'))].$$

(By the way, this sentence also serves to distinguish ω from $\omega + \zeta$.)

Exercises

59 Look up the formula Φ in Example 3.35.1. Verify that the formula $\exists X(\Phi \wedge x \in X)$ also defines the set of even natural numbers in ω.

60 Cf. Exercise 58. $\Phi(x, y)$ is the monadic second-order formula

$$\exists X(\forall z(z \in X \leftrightarrow (\mathbf{r}(x, z) \vee \exists z'(z' \in X \wedge \mathbf{r}(z', z)))) \wedge y \in X).$$

1. Show that Φ defines the usual ordering $<$ of $\mathbb{N}$ in the successor structure ω^{suc}.
2. Show that the usual ordering of $\mathbb{Z}$ is not defined by Φ in ζ^{suc}.
3. Produce a monadic second-order formula that defines the ordering in both structures ω^{suc} and ζ^{suc}. (Can you find one that is $\Pi^1_1(mon)$ and one that is $\Sigma^1_1(mon)$?)

61 Formulate and prove a counterpart of Lemma 2.4 (page 12) for monadic second-order languages.

A *graph* is a model (G, S), where S is a binary relation on G that is symmetric. A graph is called *connected* if for all $a, b \in G$ there is a finite sequence $a_1 = a, \ldots, a_n = b$ such that for all i, $1 \leq i < n$: $a_i S a_{i+1}$.

The notion of *connectedness* is $\Pi^1_1(mon)$ on the class of graphs: G is connected iff for all $a, b \in G$ and all $U \subset G$: if $a \in U$ and U is closed under neighbourship, then $b \in U$. However, the connected graph ω^{nb} is elementarily equivalent to the graph $(\omega + \zeta)^{nb}$, which is not connected. Thus, the first claim of the next proposition.

3.36 Proposition. *Connectedness is not first-order definable on the class of graphs. In fact, it is not $\Sigma^1_1(mon)$ on the class of finite graphs.*

62 Show that no first-order sentence σ exists such that for all *finite* graphs $\mathcal{A}$: $\mathcal{A} \models \sigma$ iff $\mathcal{A}$ is connected.
Hint. Use models of the form C_m and $C_m + C_m$, respectively (cf. Exercise 57).

Connectedness *is* Σ^1_1 on finite graphs: S is connected iff there exists a strict partial ordering $\prec$ with a least element, such that if y is an immediate $\prec$-successor of x, then xRy.

By König's Lemma B.7 (page 115), a finitely branching tree is finite iff all of its branches are finite. Thus, finiteness is $\Pi^1_1(mon)$ on the class of finitely branching trees. However:

3.37 Proposition. *Finiteness is not $\Sigma^1_1(mon)$ on the class of binary trees.* $\dashv$

Finally, here is an example of a partition argument.

3.38 Proposition. *Assume that $L = \{<\}$. Every L-sentence with a well-ordered model has a well-ordered model of type $< \omega^\omega$.*

Proof. Suppose that $\alpha = (A, <)$ is a well-ordering. It suffices to show that for every n, α has a well-ordered n-equivalent of type $< \omega^\omega$. Fix n. By the Downward Löwenheim-Skolem Theorem, there is no loss of generality in assuming that A is countable. Apply induction with respect to the order type of α.

If α has only one element, then α itself is the required n-equivalent. (For, $1 < \omega^\omega$.)

Next, suppose that $\alpha = \beta + 1$. Then by induction hypothesis, β has such an n equivalent β', and (by Lemma 3.12.4) $\beta' + 1 \equiv^n \beta + 1 = \alpha$ is the required equivalent. (Note that if $\beta < \omega^\omega$, then $\beta + 1 < \omega^\omega$.)

Finally, assume that α has a limit type. Let $a_0 \in \alpha$ be the least element of α. Since α is countable, there is a countable sequence $a_0 < a_1 < a_2 < \cdots$ that is unbounded in α. For $a, b \in \alpha$, $a < b$, the notation $[a, b)$ is used for the submodel of α with universe $\{x \in \alpha \mid a \leq x < b\}$. For $i < j$, let $F(i, j)$ be the set of rank-n sentences true in $[a_i, a_j)$. By Lemma 3.23, you may think of F as taking finitely many values only. By Ramsey's Theorem B.8, there exist $k_0 < k_1 < k_2 < \cdots$ such that all $F(k_i, k_j)$ are the same. By induction hypothesis, there is a well-ordering $\gamma < \omega^\omega$ that is an n-equivalent of every $[a_{k_i}, a_{k_j})$. Again by induction hypothesis, let β be a well-ordering of type $< \omega^\omega$ that is n-equivalent with $[a_0, a_{k_0})$. Then (by Lemma 3.13.2)

$$\beta + \gamma \cdot \omega \equiv^n [a_0, a_{k_0}) + \sum_i [a_{k_i}, a_{k_{i+1}}) = \alpha,$$

hence $\beta + \gamma \cdot \omega$ is the required n-equivalent of α. (Note that if $\beta, \gamma < \omega^\omega$, then $\beta + \gamma \cdot \omega < \omega^\omega$.) $\dashv$

Let Ω be the well-ordering of *all* ordinals.

3.39 Corollary. $\Omega \equiv \omega^\omega$.

Proof. See Exercise 67. $\dashv$

Exercises

63 Show that the class of models with an even number of elements is $\Delta^1_1(mon)$ definable in the class of finite linear orderings, but not first-order definable.

64 Show that every $\Sigma^1_1(mon)$ sentence true of ω is also true of $\omega + \zeta$. Nevertheless: produce a set X of natural numbers such that no expansion of $\omega + \zeta$ is elementarily equivalent to (ω, X).

65 Is every $\Sigma^1_1(mon)$ sentence true of λ true of η as well?

66 Show that if $\alpha < \beta \leq \omega^\omega$, then $\alpha \not\equiv \beta$.
Hint. Use the Cantor Normal Form from Section B.4.

67 Prove Corollary 3.39.
Hint. Show that $\Omega \equiv^n \omega^\omega$ by induction with respect to n. Use Lemma 3.10 and the fact that final segments of Ω (ω^ω) have type Ω (ω^ω).

68 ♣ A linear ordering is *scattered* if it does not embed η. Let Σ be the least set of order types such that (i) $0,\, 1 \in \Sigma$, (ii) $\alpha,\, \beta \in \Sigma \Rightarrow \alpha + \beta \in \Sigma$, (iii) $\alpha \in \Sigma \Rightarrow \alpha \cdot \omega, \alpha \cdot \omega^\star \in \Sigma$. Show that every ordering in Σ is scattered, and that every sentence with a scattered model has a model in Σ.
Hint. Use the technique of the proof of Proposition 3.38. Suppose that a certain first-order sentence of quantifier rank q is true in the scattered model $(A, <)$. Without loss of generality, assume that A is countable. Identify every submodel of $(A, <)$ with its universe. For $a, c \in A$, write $a \sim c$ in case that (i) $a < c$ and for all a', c' such that $a \leq a' < c' \leq c$,

$$(a', c') := \{b \in A \mid a' < b < c'\}$$

has a q-equivalent in Σ, or (ii) $c < a$ and a similar statement holds, or (iii) $a = c$. Then $\sim$ is an equivalence. Clearly, if $a \sim c$ and $a < b < c$, then $a \sim b$. Thus, A is an ordered sum of equivalence classes $\sum_{i \in I} C_i$, where I is a certain linear ordering.

Show that the order type of I is dense. Since $(A, <)$ is scattered, conclude that I is a singleton; i.e.: A is the only equivalence class.

Finally, show that A itself has a q-equivalent in Σ. If A has no greatest element, choose $a_0 < a_1 < a_2 < \cdots$ cofinal in A and apply Ramsey's theorem to see that $\{c \in A \mid a_0 < c\}$ has a q-equivalent in Σ. Do this also for $\{c \in A \mid c < a_0\}$, by choosing, if necessary, $b_0 = a_0 > b_1 > b_2 > \cdots$ co-initial in A.

3.3.3 (Exercises about) Theories

As always, 'sentence' means 'first-order sentence', unless the contrary is explicitly stated.

3.40 Theories. A *theory* is just a set of sentences. Theories are *equivalent* is they have the same models. A theory is *complete* if all its models are pairwise equivalent.

69 Show that theories are equivalent iff they have the same logical consequences. (Definition 1.8, page 5.)

Because of this result, a theory Σ is usually identified with the set $\{\varphi \mid \Sigma \models \varphi\}$ of sentences that logically follow from it. In that way, for a theory Σ and a sentence φ the expression '$\Sigma \models \varphi$' becomes tantamount with '$\varphi \in \Sigma$'.

3.41 Examples.

1. The empty set of sentences — equivalently: the set of logically valid sentences — is the smallest theory. *Every* model is a model of this theory.
2. There are also theories *without* models, for instance, $\{\exists x \neg x = x\}$. These theories are trivially complete, and equivalent to the theory of all sentences: the largest theory.
3. If $\mathcal{A}$ is a model, then the *theory of* $\mathcal{A}$, which is the set of all sentences true in $\mathcal{A}$, a complete theory. Notation: $Th(\mathcal{A})$.
4. If K is a class of models, then $Th(K) := \bigcap_{\mathcal{A} \in K} Th(\mathcal{A})$, the *theory of* K, is a theory.

70 Show that for every two models $\mathcal{A}$ and $\mathcal{B}$, the following conditions are equivalent:

1. $\mathcal{A} \equiv \mathcal{B}$,
2. $Th(\mathcal{A}) = Th(\mathcal{B})$,
3. $Th(\mathcal{A}) \subset Th(\mathcal{B})$,
4. $\mathcal{B} \models Th(\mathcal{A})$.

Hint. For the implication 3 $\Rightarrow$ 2, note that if $\varphi \notin Th(\mathcal{A})$, then $\neg\varphi \in Th(\mathcal{A})$.

71 Assume that Σ is a theory. Show that the following conditions are equivalent.

1. Σ has a model,
2. not every sentence follows logically from Σ,
3. there is no sentence φ such that both $\Sigma \models \varphi$ and $\Sigma \models \neg\varphi$.

Show that the following conditions are equivalent.

1. There is a model the theory of which is equivalent with Σ,
2. Σ has a model and is complete,
3. for every sentence φ it holds that either $\Sigma \models \varphi$, or $\Sigma \models \neg\varphi$,
4. Σ has a model, and every theory containing Σ that has a model is equivalent with Σ.

Dense linear orderings without endpoints. LO is the first-order theory (in the vocabulary $\{<\}$) of linear orderings. (This needs only four sentences.) DO is the first-order theory of dense linear orderings without endpoints. (Three more sentences.)

Prominent models of DO are η and λ.

72 Give some other models of DO. Give *infinitely many* models of DO.

73 Show that for every sentence φ: $\eta \models \varphi$ iff $DO \models \varphi$.

3.42 Axiomatizations. The set of sentences Σ *axiomatizes* the model $\mathcal{A}$ (or its theory $Th(\mathcal{A})$) if for every sentence φ: $\mathcal{A} \models \varphi$ iff $\Sigma \models \varphi$; equivalently: if Σ and $Th(\mathcal{A})$ are equivalent.

The set of sentences Σ *axiomatizes* the class of models K (or its corresponding theory $Th(K)$) if for every sentence φ: φ is true in every model in K iff $\Sigma \models \varphi$; equivalently: if Σ and $Th(K)$ are equivalent.

Thus, Exercise 73 shows that η is axiomatized by DO; furthermore, $\emptyset$ axiomatizes the class of all models (relative a given vocabulary), and the set of all sentences axiomatizes the empty class of models.

74 Show that no *finite* axiomatization is possible for the linear ordering $\omega \mid \omega^\star$.

Hint. Suppose that σ is the conjunction of all sentences in such a hypothetical finite axiomatization. Let σ have quantifier rank n and let $m := 2^n$. Then (Lemma 3.12.2, page 25) $\boldsymbol{m-1}$ is a model of σ. However, the sentence that expresses the existence of at least m elements is not true in this model, whereas it is true in $\omega + \omega^\star$.

Finite linear orderings. FLO is the set of first-order sentences extending LO with sentences expressing that: there is a first element, a last element, and every element that has a successor (predecessor) also has an immediate successor (immediate predecessor).

3.43 Lemma. *Every infinite model of* FLO *has an order type of the form* $\omega + \zeta \cdot \alpha + \omega^\star$.

Proof. Assume that $\mathcal{A} = (A, <)$ is an infinite model of FLO. $\mathcal{A}$ has an initial of order type ω: there is a first element 0; this is not the only element, so it has an immediate successor 1, which in turn has an immediate successor 2, etc. Similarly, $\mathcal{A}$ has a tail $-1 > -2 > -3 > \cdots$ of order type $\omega^\star$.

If this all of $\mathcal{A}$, then $\mathcal{A}$ has order type $\omega + \zeta \cdot \alpha + \omega^\star$ where $\alpha = 0$. If not, there exists some $a \in A$ different from these elements. So

$$0 < 1 < 2 < \cdots < a < \cdots - 3 < -2 < -1.$$

Since a has successors such as -1, it has an immediate successor $a + 1$; $a + 1$ in turn has an immediate successor $a + 2$ etc.; similarly, a has an immediate predecessor $a-1$ etc. This series of successors and predecessors form an interval $\ldots, a-2, a-1, a, a+1, a+2, \ldots$ of type ζ.

The collection of these intervals of type ζ itself is ordered in some type α. It follows that $\mathcal{A}$ has type $\omega + \zeta \cdot \alpha + \omega^\star$. $\dashv$

75 Show that *FLO* axiomatizes the class of finite linear orderings.

76 Show that $FLO \cup \{M_m \mid m \in \mathbb{N}^+\}$ (where M_m is a sentence expressing the existence of at least m elements) axiomatizes the linear ordering $\omega + \omega^\star$.

77 Suppose that Σ has arbitrarily large, finite linear orderings as models. I.e.: for every $n \in \mathbb{N}$ there exists $m \in \mathbb{N}$ such that $n \leq m$ and such that $\mathbf{m}$ is a model of Σ. Show that $\omega + \omega^\star$ also is a model of Σ.

78 Give a $\Pi^1_1(mon)$-sentence ε with the property that a linear ordering satisfies $FLO \cup \{\varepsilon\}$ iff it is finite.

79 Construct a finite axiomatisation for ω. Do this in such a way that every model of your axioms in which the monadic second-order sentence from Example 3.35.2 holds, is isomorphic with ω.
Hint. Use the method of Lemma 3.43 and Exercise 75.

80 Produce a finite axiomatization for ζ. Give a monadic second-order sentence that is true of ζ such that every model of your axioms in which this sentence holds true is actually isomorphic with ζ.

81 Show that the following two conditions are equivalent.

1. $\mathcal{A}$ has a finite axiomatization,
2. there exists a natural number n such that for every model $\mathcal{B}$: $\mathcal{B} \equiv^n \mathcal{A} \Rightarrow \mathcal{B} \equiv \mathcal{A}$.

82 ♣ Show that if the linear orderings α and β are finitely axiomatizable, then so are $\alpha^\star$, $1+\alpha$, $\alpha+1$ and $\alpha+1+\beta$. Give an example showing that $\alpha+\beta$ is not necessarily finitely axiomatizable.

Successor relations. The infinite set *SUCC* consists of the following sentences:

$$\forall x \exists y (\mathbf{r}(x,y) \wedge \forall z (\mathbf{r}(x,z) \rightarrow z = y))$$

and

$$\forall x \exists y (\mathbf{r}(y,x) \wedge \forall z (\mathbf{r}(z,x) \rightarrow z = y)),$$

as well as

(1) $\neg \exists x_1 \mathbf{r}(x_1, x_1)$
(2) $\neg \exists x_1 \exists x_2 (\mathbf{r}(x_1, x_2) \wedge \mathbf{r}(x_2, x_1))$
(3) $\neg \exists x_1 \exists x_2 \exists x_3 (\mathbf{r}(x_1, x_2) \wedge \mathbf{r}(x_2, x_3) \wedge \mathbf{r}(x_3, x_1))$
(4) $\ldots$

83 Show that every model of *SUCC* is of the form $(\zeta \cdot \alpha)^{suc}$ for a suitable α.

Categoricity. A theory is *categorical in* a cardinal if each of its models of that cardinal are pairwise isomorphic.

Examples. DO is categorical in $\aleph_0$ (Corollary 3.49) but in no uncountable cardinal. $SUCC$ is categorical in every uncountable cardinal but not in $\aleph_0$.

84 Show that $SUCC$ has infinitely many non-isomorphic countable models.

85 ♣ Show that $SUCC$ is categorical in every uncountable cardinal.

86 Show that every sentence true of ζ^{suc} also has a finite model. (In particular: ζ^{suc} is not finitely axiomatizable.)

Łoś' conjecture. *If a theory in a countable vocabulary is categorical in one uncountable cardinal, then it is categorical in every uncountable cardinal.*

The following deep, thirty year old result forms the germ of the field known as classification theory.

Morley's Theorem. *The Łoś' conjecture is correct.*

87 Produce a theory that has models of every infinite cardinality which is categorical in every infinite cardinal.

3.44 Łoś-Vaught test. *Every theory in a countable vocabulary that is categorical in some infinite cardinal and has some infinite model is complete.*

88 Prove the Łos-Vaught test.
Hint. Use the Downward and Upward Löwenheim-Skolem-Tarski Theorems 2.13 and 4.10.

3.4 The Infinite Game

3.45 Definition. In the *infinite* Ehrenfeucht game $E(\mathcal{A}, \mathcal{B}, \infty)$ on $\mathcal{A}$ and $\mathcal{B}$, there is no bound on the number of moves; Di and Sy alternate in making an ω-sequence of moves each, and win and loss are determined (almost) as before: Sy *wins* if at each finite stage of the play, the moves made so far constitute a local isomorphism between the models.

$\mathcal{A}$ and $\mathcal{B}$ are *partially isomorphic* if Sy has a winning strategy for $E(\mathcal{A}, \mathcal{B}, \infty)$.

3.46 Examples.

1. η and λ are partially isomorphic. Better still:
2. Every two dense linear orderings without endpoints are partially isomorphic.
3. No well-ordering is partially isomorphic with a non-well-ordering. (Let Di play an infinite descending sequence in the non-well-ordering. Note that this argument also works for the 2-pebble game.)
4. Well-orderings of different type are not partially isomorphic. (To begin with, Di plays the element a of the larger one such that $a\!\downarrow$ has

the type of the smaller one. Subsequently, Di can always counter a move b of Sy with a move c such that $c\downarrow$ and $b\downarrow$ have the same type. Eventually, she must out-play Sy. For this argument, again 2 pebbles suffice.)

3.47 Proposition. *In every infinite Ehrenfeucht game, exactly one of the players has a winning strategy.*

Proof. Note that if Di wins a play, this fact becomes apparent after finitely many moves already: the game is *open*. See Section B.9. ⊣

The following important theorem has an extremely simple proof. (Compare Proposition 3.9.)

3.48 Theorem. *Countable partially isomorphic models are isomorphic.*

Proof. Let Di enumerate all elements of the two models and let Sy use his winning strategy. The resulting play constitutes the isomorphism we are looking for. ⊣

Cantor's characterization of the ordering η of the rationals is an immediate corollary. The proof of Theorem 3.48 is an abstract version of the usual back-and-forth proof for the Cantor result.

3.49 Corollary. *The linear ordering η is (up to isomorphism) the only countable dense linear ordering without endpoints.* ⊣

Of course, the game E^h has an infinite version as well, with its corresponding notion of *partial homomorphism*. Theorem 3.48 now modifies to:

3.50 Proposition. *If the countable models $\mathcal{A}$ and $\mathcal{B}$ are partially homomorphic, then there is a homomorphism from $\mathcal{A}$ onto $\mathcal{B}$.* ⊣

Similarly:

3.51 Proposition. *If* Sy *has a winning strategy in the infinite $E^{\mathbf{r}\text{-pos}}$-game on the countable $L \cup \{\mathbf{r}\}$-models $\mathcal{A}$ and $\mathcal{B}$, then $\mathcal{A} \mid L \cong \mathcal{B} \mid L$ and the isomorphism is an $\mathbf{r}$-homomorphism.*

To explain the logical meaning of the infinite game, you need the notion of an *infinitary* formula. This is obtained by modifying the definition of first-order formula, admitting conjunctions and disjunctions of arbitrarily many formulas. I.e., if L is a vocabulary, the class $L_{\infty\omega}$ of *infinitary L-formulas* is obtained by adding to Definition 1.2 (page 3) the clause

4. if Φ is an arbitrary set of formulas, then $\bigwedge \Phi$ and $\bigvee \Phi$ are formulas.

The semantics of such infinitary formulas is rather obvious: the formula $\bigwedge \Phi$ ($\bigvee \Phi$) is satisfied by the assignment α in the model $\mathcal{A}$ iff *every* (*some*) $\varphi \in \Phi$ is. (This implies that $\bigwedge \emptyset$ is always satisfied whereas $\bigvee \emptyset$ never is, and that $\bigwedge\{\varphi\}$ and $\bigvee\{\varphi\}$ are logically equivalent with φ.)

Equivalence with respect to infinitary sentences is denoted by $\equiv_{\infty\omega}$. (In this notation, the ∞ signifies that arbitrary con- and disjunctions are admitted; the ω indicates that quantification still is restricted to finitely many variables at the same time.)

The following proposition explains that the infinite game is not just the limit of the finite games.

Recall the monotone operator

$$\Gamma(X) =$$
$$\{h \mid \forall a \in A\, \exists b \in B\, (h \cup \{(a,b)\} \in X) \wedge \forall b \in B\, \exists a \in A\, (h \cup \{(a,b)\} \in X)\}$$

defined by (1) on page 28. The finite stages $\Gamma{\downarrow}n$ of the downward hierarchy of this operator were relevant to the finite game. Let W be the set of relations $\{(a_1,b_1),\dots,(a_n,b_n)\}$ such that Sy has a winning strategy for the infinite game on $(\mathcal{A},a_1,\dots,a_n)$ and $(\mathcal{B},b_1,\dots,b_n)$. Let EQ be the set of relations $\{(a_1,b_1),\dots,(a_n,b_n)\}$ such that $(\mathcal{A},a_1,\dots,a_n) \equiv_{\infty\omega} (\mathcal{B},b_1,\dots,b_n)$.

Part of the second equality of the following result says that Sy has a winning strategy for the infinite game between two models iff they cannot be distinguished using infinitary sentences.

3.52 Proposition. $\Gamma{\downarrow} = W = EQ$.

Proof. In view of Lemma B.3, it suffices to show that both W and EQ are co-inductive post-fixed points. Those interested only in the equality $W = EQ$ (the infinitary analogue of Theorem 3.18) are referred to Exercise 91.

It is easy to see that W is a post-fixed point. Co-inductiveness: assume that X is a set of local isomorphisms such that $X \subset \Gamma(X)$. Suppose that $h \in X$. To see that $h \in W$, the strategy of Sy is taking care that for every position $\{(a_1,b_1),\dots,(a_n,b_n)\}$ visited in the playing of the game, he has $h \cup \{(a_1,b_1),\dots,(a_n,b_n)\} \in X$. If he succeeds in doing so, he wins. That he can succeed follows from X being a post-fixed point.

EQ is also co-inductive. For, assume that $X \subset \Gamma(X)$. It follows by induction on sentences (keeping h variable) that every

$$h := \{(a_1,b_1),\dots,(a_n,b_n)\} \in X$$

satisfies

$$(\mathcal{A},a_1,\dots,a_n) \equiv_{\infty\omega} (\mathcal{B},b_1,\dots,b_n).$$

That EQ is a post-fixed point can be shown using the method of proof of Theorem 3.18. Note that you do not need Lemma 3.23, since infinite conjunctions are allowed. ⊣

Exercises

89 Let C be a (countably) infinite set of constant symbols. Show that the infinitary sentence $\forall x \bigvee_{\mathbf{c} \in C} x = \mathbf{c}$ does not have a first-order equivalent. *Hint.* Use Exercise 16.

90 Suppose that $\mathcal{A} = (A, <)$ is a well-ordering. Recursively define, for $a \in A$, the infinitary formula φ_a as $\forall y(y < x \leftrightarrow \bigvee_{b<a} \varphi_b(y))$. (If you encounter problems with substituting into an infinitary formula, you might use $\forall y(y < x \leftrightarrow \exists x(y = x \wedge \bigvee_{b<a} \varphi_b))$. Thus, every φ_a uses two variables x and y; exactly one occurrence of x is free.) Let $\Phi_{\mathcal{A}}$ be the infinitary sentence $\forall x \bigvee_{a\in A} \varphi_a \wedge \bigwedge_{a\in A} \exists x \varphi_a$.

Show the following:

1. $(A, <) \models \varphi_a[b]$ iff $b = a$,
2. a linear ordering satisfies $\Phi_{\mathcal{A}}$ iff it is an isomorph of $\mathcal{A}$.

91 Give a direct proof that $W = EQ$.

Sketch. ($\Rightarrow$) By induction with respect to the infinitary sentence φ show that *if* $\mathsf{Sy}(\mathcal{A}, \mathcal{B}, \infty)$, *then* $\mathcal{A} \models \varphi \Leftrightarrow \mathcal{B} \models \varphi$. For atomic sentences, this follows from the fact that $\mathsf{Sy}(\mathcal{A}, \mathcal{B}, \infty)$ implies $\mathsf{Sy}(\mathcal{A}, \mathcal{B}, 0)$. The induction steps for the connectives are completely trivial. Finally, assume that $\mathcal{A} \models \exists x \varphi(x)$. Let $a \in A$ be such that $(\mathcal{A}, a) \models \varphi(\mathbf{c})$, where the new constant symbol $\mathbf{c}$ is interpreted as a in the expansion $(\mathcal{A}, a)$. Consider a as a first move of Di in the infinite game. Let an answer $b \in B$ be determined by some winning strategy. Since the game is infinite, obviously we have that $\mathsf{Sy}((\mathcal{A}, a), (\mathcal{B}, b), \infty)$. Therefore, by induction hypothesis applied to $\varphi(\mathbf{c})$, we have that $(\mathcal{B}, b) \models \varphi(\mathbf{c})$. It follows that $\mathcal{B} \models \exists x \varphi$.

($\Leftarrow$) Assume that $\mathcal{A}$ and $\mathcal{B}$ have the same true infinitary sentences. The strategy of Sy consists in playing in such a way that after his n-th move a position $\{(a_1, b_1), \ldots, (a_n, b_n)\}$ is obtained for which

$$(\mathcal{A}, a_1, \ldots, a_n) \equiv_{\infty\omega} (\mathcal{B}, b_1, \ldots, b_n). \tag{2}$$

If Sy can take care of this, he obviously wins. So, let us check that this is indeed possible. Suppose that Sy has succeeded in maintaining this condition up to and including the n-th pair of moves as in (2). Suppose that Di plays $a_{n+1} \in A$. Consider the set $\Phi := \{\varphi \mid (\mathcal{A}, a_1, \ldots, a_n) \models \varphi[a_{n+1}]\}$ of formulas satisfied by this move. Then

$$(\mathcal{A}, a_1, \ldots, a_n) \models \bigwedge \Phi[a_{n+1}]$$

and hence

$$(\mathcal{A}, a_1, \ldots, a_n) \models \exists x_{n+1} \bigwedge \Phi.$$

By assumption, $(\mathcal{B}, a_1, \ldots, b_n) \models \exists x_{n+1} \bigwedge \Phi$. Let Sy choose $b_{n+1} \in B$ satisfying $\bigwedge \Phi$ in the expanded $\mathcal{B}$. Then obviously,

$$(\mathcal{A}, a_1, \ldots, a_{n+1}) \equiv_{\infty\omega} (\mathcal{B}, b_1, \ldots, b_{n+1}),$$

as desired.

92 Define the quantifier rank of infinitary formulas. Connect $\Gamma{\downarrow}\alpha$ to quantifier rank-α equivalence. Can you concoct a matching notion of α-game?

When $\mathcal{A} = \mathcal{B}$, the closure ordinal of Γ is the *Scott rank* of $\mathcal{A}$.

93 Show that the Scott rank of the linear ordering ω equals ω. Give an example of a model with Scott rank $> \omega$.

3.53 Characteristics. Let $\mathcal{A}$ be a model. For every finite sequence $\vec{a} = (a_1, \ldots, a_n)$ from A and every ordinal α, define the infinitary quantifier rank-α formula $[\![\vec{a}]\!]^\alpha$ in the free variables $x_1, \ldots, x_n$, the *α-characteristic* of $\vec{a}$ in $\mathcal{A}$, as follows.

$[\![\vec{a}]\!]^0$ is the conjunction of all formulas in $x_1, \ldots, x_n$ satisfied by $\vec{a}$ in $\mathcal{A}$ that are either atomic or negations of atomic formulas.

$$[\![\vec{a}]\!]^\gamma = \bigwedge_{\xi<\gamma} [\![\vec{a}]\!]^\xi, \text{ when } \gamma \text{ is a limit.}$$

$$[\![\vec{a}]\!]^{\alpha+1} = \bigwedge_{b\in A} \exists x_{n+1} [\![\vec{a}, b]\!]^\alpha \wedge \forall x_{n+1} \bigvee_{b\in A} [\![\vec{a}, b]\!]^\alpha.$$

94 Show the following:

1. for all α, $\mathcal{A} \models [\![\vec{a}]\!]^\alpha[\vec{a}]$,
2. $\mathcal{B} \models [\![\vec{a}]\!]^\alpha[\vec{b}]$ iff $(\mathcal{A}, \vec{a})$ and $(\mathcal{B}, \vec{b})$ satisfy the same quantifier rank $\leq \alpha$ formulas, iff $[\![\vec{b}]\!]^\alpha = [\![\vec{a}]\!]^\alpha$.

If α is the Scott rank of $\mathcal{A}$, then $[\![(\,)]\!]^\alpha \wedge \bigwedge_{\vec{a}} \forall \vec{x}([\![\vec{a}]\!]^\alpha \rightarrow [\![\vec{a}]\!]^{\alpha+1})$ is the *Scott sentence* of $\mathcal{A}$.

The language $L_{\omega_1\omega}$ is the restriction of $L_{\infty\omega}$ that allows conjunctions and disjunctions over countable sets of formulas only. Note that the Scott sentence of a countable model belongs to this language.

95 Show that the Scott sentence of a model axiomatizes its infinitary theory.

Bibliographic Remarks

Ehrenfeucht's game is from Ehrenfeucht 1961, Fraïssé's formulation is from Fraïssé 1954.

Proposition 3.24 and its corollary are due to Immerman and Kozen 1989.

Theorem 3.34 is due to Fagin 1974. Immerman (see Immerman 1987a, 1987b, 1989) has made a detailed study of the relations between complexity levels and definability in logical formalisms of several type.

Proposition 3.36 is from Fagin 1975. Also, see Gaifman and Vardi 1985, Fagin et al. 1992 and Ajtai and Fagin 1990. For the development of finite-model theory, see Fagin 1990. The fact that connectedness is Σ_1^1 on finite graphs was noted by Fagin. He also showed that any example of a property of finite models that is Σ_1^1 but not Π_1^1 would entail that P $\neq$ NP, and if such a property exists at all, then 3-colorability (which is NP-complete) must be an example.

Morley's Theorem (1965) is the starting point for what is known as *classification theory*. The result in Exercise 91 (*Karp's Theorem*) is from Karp 1965. Much more on infinitary logic is in Keisler 1971 and Barwise 1975.

Everything you want to know about linear orderings is in Rosenstein 1982.

For a growing source of publications on finite-model theory, see the list FMT 1995, *References for Finite Model Theory*. For a recent monograph on this subject, see Ebbinghaus and Flum 1995.

4

Constructing Models

In Chapters 2 and 3, several relations between models were introduced by means of which they could be compared. In the present chapter you will be introduced to several ways of constructing models. The basic instrument of first-order model theory is the Compactness Theorem of Section 4.1. This tool is essential for most applications of the diagram method of Section 4.2. The ultraproducts of Section 4.3 look like magic, although their use in model theory is restricted. A refinement of the compactness proof produces the Omitting Types Theorem of Section 4.4. Compactness also is responsible for the saturated models of Section 4.5 that are applied in Section 4.7.

4.1 Compactness

The Compactness Theorem is one of the main tools in constructing models for sets of first-order sentences.

A set of sentences is *satisfiable* if it has a model; it is *finitely satisfiable* if all its finite subsets are satisfiable.

4.1 Compactness Theorem (version 1). *Every finitely satisfiable set of first-order sentences is satisfiable.*

Second-order logic is incompact. It is easy to write down a $\Pi^1_1(mon)$-sentence Π whose only model (up to isomorphism) is the linear ordering ω. (See Exercise 79.) Let **c** be an individual constant and consider the set $\{\Phi\} \cup \{\exists n \mid n \in \mathbb{N}\}$, where $\exists n$ expresses that **c** has at least n predecessors. This set is finitely satisfiable but not satisfiable. Also, infinitary logic is incompact: consider the set $\{\bigvee_n \neg\exists n\} \cup \{\exists n \mid n \in \mathbb{N}\}$.

Here are two variations on the Compactness Theorem that are often needed.

Compactness Theorem (version 2). *If* $\Sigma \models \varphi$, *then for some finite* $\Delta \subset \Sigma$, $\Delta \models \varphi$.

Proof. If φ does not logically follow from some finite $\Delta \subset \Sigma$, then the set

$\Sigma \cup \{\neg\varphi\}$ is finitely satisfiable and therefore, by Compactness, satisfiable. So $\Sigma \not\models \varphi$. $\dashv$

A set Γ of formulas is *satisfiable* if there are a model and an assignment that satisfies every formula from Γ in the model. A set of formulas is *finitely satisfiable* if each of its finite subsets is satisfiable.

The following result says that Compactness also applies to sets of formulas.

Compactness Theorem (version 3). *Every finitely satisfiable set of first-order formulas is satisfiable.*

Proof. Let Γ be a finitely satisfiable set of L-formulas. The problem is reduced to Compactness for sentences, Theorem 4.1. Choose an injection h of the set of variables into some set C of new constant symbols. In the formulas of Γ, replace every occurrence of a free variable x by its h-image. This results in a finitely satisfiable set Γ' of $L \cup C$-*sentences.* By Compactness, Γ' has a model. But then the assignment that maps a variable x to the interpretation of $h(x)$ in this model satisfies the original formulas of Γ. $\dashv$

The remaining part of this section deals with proving the Compactness Theorem.

A set of sentences is *maximally finitely satisfiable* if it is finitely satisfiable but has no proper extension in the *same* vocabulary that enjoys this property.

The germ of first-order Compactness is propositional Compactness; and this is more or less the content of the following lemma.

4.2 Lemma. *Every finitely satisfiable set of sentences can be extended to a maximal finitely satisfiable set of sentences.*

Proof. Given the properties of the required object, it appears sensible to apply Zorn's Lemma (see Section B.6) to the collection of finitely satisfiable extensions of the given set, partially ordered by the inclusion relation. Indeed, the union of a chain of such extensions will again be finitely satisfiable. $\dashv$

The interest in maximal finite satisfiability derives from the following lemma.

4.3 Lemma. *Assume that Γ is a maximal finitely satisfiable set of sentences. Then for all sentences φ, ψ and sets of sentences Δ:*

1. *if $\Delta \subset \Gamma$ is finite and $\Delta \models \varphi$, then $\varphi \in \Gamma$,*
2. *$\varphi \wedge \psi \in \Gamma$ iff $\varphi, \psi \in \Gamma$,*
3. *$\neg\varphi \in \Gamma$ iff $\varphi \notin \Gamma$.*

Proof. 1. Assume that the sentence φ logically follows from the finite

$\Delta \subset \Gamma$. To see that $\varphi \in \Gamma$, by maximality it suffices to show that $\Gamma \cup \{\varphi\}$ is finitely satisfiable. So let $\Sigma \subset \Gamma$ be finite; it must be shown that $\Sigma \cup \{\varphi\}$ is satisfiable. Now $\Delta \cup \Sigma \subset \Gamma$ is finite and hence has a model $\mathcal{A}$. Since $\Delta \models \varphi$, $\mathcal{A}$ is a model of φ as well. Thus, $\Sigma \cup \{\varphi\}$ is satisfiable.
2. Immediate from 1.
3. Exercise 96. ⊣

Recall that (Definition 3.2) the vocabulary L' *simply extends* L if $L \subset L'$ and $L' - L$ consists of constant symbols only.

4.4 Lemma. *Every finitely satisfiable set Σ of L-sentences can be extended to a finitely satisfiable set Σ' of sentences in a vocabulary L' simply extending L such that*

if $\exists x \varphi(x) \in \Sigma$, then for some individual constant $\mathbf{c} \in L'$: $\varphi(\mathbf{c}) \in \Sigma'$.

Proof. Add a new constant $\mathbf{c}_\varphi$ for each existential L-sentence $\exists x \varphi(x)$; $\mathbf{c}_\varphi$ is called a *witness* for φ. Define $\Sigma' := \Sigma \cup \{\varphi(\mathbf{c}_\varphi) \mid \exists x \varphi(x) \in \Sigma\}$. See Exercise 97 for details. ⊣

The interest now is a property that strengthens that of Lemma 4.4. A set of sentences $\Sigma^\star$ has the *Henkin property* if, for every existential sentence $\exists x \varphi(x)$ in $\Sigma^\star$, there exists a constant symbol $\mathbf{c}$ such that $\varphi(\mathbf{c}) \in \Sigma^\star$. Note the difference between the Henkin property and the condition from Lemma 4.4.

Alternating the two ways of extending a finitely satisfiable set countably many times, you will eventually obtain a set that is both finitely satisfiable and Henkin:

4.5 Corollary. *Let L be a vocabulary. There exists a vocabulary $L^\star \supset L$ simply extending L such that every finitely satisfiable set of L-sentences can be extended to a maximal finitely satisfiable set of $L^\star$-sentences that has the Henkin property.*

Proof. Left as Exercise 98. ⊣

Now, Compactness follows once you've proved:

4.6 Lemma. *Every maximal finitely satisfiable set with the Henkin property has a model.*

Proof. Let Γ be maximal finitely satisfiable and Henkin. Here is the basic Henkin construction of the *canonical model* $\mathcal{A}$ for Γ.

Define the relation $\sim$ on the set of variable-free terms by

$$s \sim t :\equiv (s = t) \in \Gamma.$$

From Lemma 4.3.1 it follows that this is an equivalence. The universe A of the required model is the set of equivalence classes of variable-free terms $|t| := \{s \mid s \sim t\}$. This universe is provided with structure interpreting the non-logical symbols of the vocabulary of Γ.

For every individual constant $\mathbf{c}$, let its interpretation be $\mathbf{c}^{\mathcal{A}} := |\mathbf{c}|$.

If $\mathbf{f}$ is an n-ary function symbol, define its interpretation $\mathbf{f}^{\mathcal{A}} : A^n \to A$ by $\mathbf{f}^{\mathcal{A}}(|t_1|, \ldots, |t_n|) := |\mathbf{f}(t_1, \ldots, t_n)|$. Note that this definition does not depend on the representatives $t_1, \ldots, t_n$ of the equivalence classes $|t_1|, \ldots, |t_n|$. (This again uses Lemma 4.3.1.) You can now evaluate terms in the structure $\mathcal{A}$ defined so far.

Claim 1. For every variable-free term t: $t^{\mathcal{A}} = |t|$.
Proof. This is Exercise 99.

Finally, relation symbols are interpreted as follows. For n-ary $\mathbf{r}$, define its interpretation $\mathbf{r}^{\mathcal{A}}$ by $\mathbf{r}^{\mathcal{A}}(|t_1|, \ldots, |t_n|) :\equiv \mathbf{r}(t_1, \ldots, t_n) \in \Gamma$. Again, this definition is representative independent by Lemma 4.3.1. The required result follows from the next

Claim 2. For every sentence φ: $\mathcal{A} \models \varphi \Leftrightarrow \varphi \in \Gamma$.
Proof. Induction with respect to the number of logical symbols in φ. For atomic φ, the result is immediate from the way relations are defined and from Claim 1. The induction proceeds smoothly through the connectives: this is what Lemma 4.3.2/3 is good for. Here follows a quantifier step. If $\mathcal{A} \models \exists x \varphi(x)$, then for some $a \in A$ we have $\mathcal{A} \models \varphi[a]$. For some variable free term t, $a = |t| = t^{\mathcal{A}}$. By Exercise 4, $\mathcal{A} \models \varphi(t)$. Since $\varphi(t)$ contains one logical symbol less than $\exists x \varphi$, the induction hypothesis applies, and you obtain that $\varphi(t) \in \Gamma$. However (Exercise 6), $\varphi(t) \models \exists x \varphi(x)$; and so (Lemma 4.3.1) $\exists x \varphi \in \Gamma$. Conversely, assume that $\exists x \varphi(x) \in \Gamma$. Then (since Γ is Henkin) for some $\mathbf{c}$ we have $\varphi(\mathbf{c}) \in \Gamma$ as well. By induction hypothesis, $\mathcal{A} \models \varphi(\mathbf{c})$. Thus, $\mathcal{A} \models \exists x \varphi$. $\dashv$

Exercises

96 Prove Lemma 4.3.3.
Hint. If $\varphi \notin \Gamma$ then by maximality there is a finite $\Delta \subset \Gamma$ such that $\Delta \cup \{\varphi\}$ is not satisfiable. Apply part 1.

97 Show that the set Σ' defined in the proof of Lemma 4.4 is finitely satisfiable.

98 Prove Corollary 4.5.
Hint. Construct an ascending chain of sets of sentences that starts with the given set by alternating Lemma 4.2 and 4.4. Check that the union of this chain has the required properties.

99 Verify the two claims in the proof of Lemma 4.6.
Hint. Use term induction for the first one.

100 Prove that, in the model $\mathcal{A}$ constructed in the proof of Lemma 4.6, $A = \{|\mathbf{c}| \mid \mathbf{c} \text{ an individual constant}\}$. I.e., every equivalence class contains a constant symbol; or equivalently: every element is the interpretation of a constant symbol.

101 Assume that every model of φ satisfies a sentence from Σ. Show that a finite set $\Delta \subset \Sigma$ exists such that every model of φ satisfies a sentence from Δ.

102 Assume that φ and Γ have the same models. Show that a finite $\Delta \subset \Gamma$ exists such that φ and Δ have the same models.

103 Suppose that the sets of sentences Σ and Γ are such that every model is either a model of Σ or a model of Γ. Show that a finite $\Delta \subset \Sigma$ exists that has exactly the same models as Σ.

104 Suppose that Γ and Σ are sets of sentences and Σ is closed under $\neg$ and $\wedge$. The following are equivalent:

1. Any two models of Γ satisfying the same sentences from Σ are equivalent,
2. for every sentence φ there exists $\psi \in \Sigma$ such that $\Gamma \models (\varphi \leftrightarrow \psi)$.

Hint. For $1 \Rightarrow 2$, put $\Delta := \{\sigma \in \Sigma \mid \Gamma \models \varphi \to \sigma\}$ and show that $\Gamma \cup \Delta \models \varphi$.

105 Suppose that the sentence φ and the set of sentences Σ are such that for all models $\mathcal{A}$ and $\mathcal{B}$: if $\mathcal{A}$ and $\mathcal{B}$ satisfy the same sentences from Σ and $\mathcal{A} \models \varphi$, then $\mathcal{B} \models \varphi$. Show: φ is equivalent with a *boolean combination* of sentences from Σ (i.e., a sentence that can be obtained from elements of Σ using $\neg$, $\wedge$ and $\vee$).
Hint. Without loss of generality you may assume that Σ is closed under negations and disjunctions. Try $\Delta := \{\psi \in \Sigma \mid \varphi \models \psi\}$.

106 Let L be any vocabulary. Show: there is no set E of L-sentences such that for every L-model $\mathcal{A}$: $\mathcal{A} \models E$ iff $\mathcal{A}$ has a finite universe.
Hint. Construct, for $n \geq 1$, a sentence M_n such that $\mathcal{A} \models M_n$ iff A has at least n elements. Consider, for such a hypothetical set E, $E \cup \{M_1, M_2, M_3, \ldots\}$.

107 (A simple form of *Herbrand's Theorem.*) Suppose that $\varphi = \varphi(x)$ is a quantifier-free L-formula in one free variable x, where L contains at least one constant symbol. Show that the following conditions are equivalent:

1. $\models \exists x\varphi$,
2. there are finitely many variable-free terms $t_1, \ldots, t_n$ such that

$$\models \varphi(t_1) \vee \cdots \vee \varphi(t_n).$$

4.2 Diagrams

Applications of Compactness often involve *diagrams.* The diagram of a model is defined via its complete simple expansion. For the notion of a simple expansion, see Definition 3.2, page 21.

4.7 Diagrams. Let $\mathcal{A}$ be an L-model. Consider the elements of A as new individual constants. Adding them to L produces the vocabulary $L_A =$

$L \cup A$ that is a simple expansion of L. The *complete (simple) expansion* of $\mathcal{A}$ is the simple expansion of $\mathcal{A}$ into an L_A-model that has the element $a \in A$ itself as the interpretation of the new constant symbol a. The notation for the complete expansion is $(\mathcal{A}, a)_{a \in A}$.

The *elementary diagram* $ELDIAG(\mathcal{A})$ of the L-model $\mathcal{A}$ is the set of all L_A-sentences that are true in $(\mathcal{A}, a)_{a \in A}$. (Using the notation introduced by 3.41.3 we have $ELDIAG(\mathcal{A}) = Th((\mathcal{A}, a)_{a \in A})$.)

The *diagram* $DIAG(\mathcal{A})$ of the L-model $\mathcal{A}$ is the part of $ELDIAG(\mathcal{A})$ that contains atomic sentences and negations of atomic sentences.

Thus, $ELDIAG(\mathcal{A})$ is much bigger than $DIAG(\mathcal{A})$.

Note that every L_A-sentence can be written in the form $\varphi(a_1, \ldots, a_n)$ where $\varphi = \varphi(x_1, \ldots, x_n)$ is an L-formula with $x_1, \ldots, x_n$ free and $a_1, \ldots, a_n$ are new individual constants from A (Notation 1.8, page 5).

Of course, the new individual constants from L_A that are elements $a \in A$ will be interpreted in any L_A-model, not only in $(\mathcal{A}, a)_{a \in A}$. An arbitrary L_A-model $\mathcal{B}'$ can always be viewed as a simple expansion of its L-reduct $\mathcal{B} := \mathcal{B}' \mid L$ that interprets a new constant symbol $a \in A$ as $h(a) \in B$, where $h : A \to B$ is some function from A into B. Then $\mathcal{B}'$ usually is written as $(\mathcal{B}, h(a))_{a \in A}$.

The following lemma tells you in model theoretic terms when h is an (elementary) embedding, connecting it with the notion of (elementary) diagram.

4.8 Lemma. *Suppose that $\mathcal{A}$ and $\mathcal{B}$ are L-models and that $h : A \to B$. The following conditions are equivalent:*

1. *$(\mathcal{B}, h(a))_{a \in A}$ is a model of $DIAG(\mathcal{A})$ (respectively: of $ELDIAG(\mathcal{A})$),*
2. *h is an embedding (respectively: elementary embedding) of $\mathcal{A}$ in $\mathcal{B}$.*

Proof. See Exercise 108. ⊣

Lemma 4.8 has the following trivial, but extremely useful consequences.

4.9 Corollary. *Let $\mathcal{A}$ be an L-model. The L-reduct of a model of $DIAG(\mathcal{A})$ ($ELDIAG(\mathcal{A})$) is (up to isomorphism) an extension (elementary extension) of $\mathcal{A}$.*

A typical application of Corollary 4.9 is the following result.

4.10 Upward Löwenheim-Skolem-Tarski Theorem. *If $\mathcal{A}$ is an infinite L-model and $|A|$, $|L| \leq \mu$, then $\mathcal{A}$ has an elementary extension of cardinality μ.*

Proof. Choose a set C of new individual constants of power μ that is disjoint from $L \cup A$. Consider the set of sentences

$$\Gamma := ELDIAG(\mathcal{A}) \cup \{\neg \mathbf{c} = \mathbf{c}' \mid \mathbf{c}, \mathbf{c}' \in C \wedge \mathbf{c} \neq \mathbf{c}'\}.$$

Suppose that Γ has a model $(\mathcal{B}, h(a), c)_{a\in A, \mathbf{c}\in C}$. By Corollary 4.9, $\mathcal{B}$ is (up to isomorphism) an elementary extension of $\mathcal{A}$ that clearly must have cardinality $\geq \mu$. If its cardinality happens to be $> \mu$, then by the Downward Löwenheim-Skolem-Tarski Theorem 2.13, choose $\mathcal{C} \prec \mathcal{B}$ of cardinality exactly μ such that $\mathcal{A} \subset \mathcal{C}$. It follows (by 2.9.2, page 14) that $\mathcal{A} \prec \mathcal{C}$.

So it suffices to show that Γ has a model. This follows from Compactness once you can show that every finite subset has a model. Therefore, suppose that $\Gamma' \subset \Gamma$ is finite. Say, $\Gamma' = \Sigma \cup \Delta$, where $\Sigma \subset ELDIAG(\mathcal{A})$ and

$$\Delta \subset \{\neg \mathbf{c} = \mathbf{c}' \mid \mathbf{c}, \mathbf{c}' \in C \wedge \mathbf{c} \neq \mathbf{c}'\}.$$

Now, Σ does have a model: $(\mathcal{A}, a)_{a\in A}$. This model can be expanded into a model of Δ: you just have to find *different* interpretations for the *finitely many* constant symbols occurring in the inequalities of Δ. And this is unproblematic as, by assumption, $\mathcal{A}$ is infinite. ⊣

Here is a simple but nice application of the above material.

4.11 Universal Formulas. A *universal* formula is one of the form $\forall x_1 \cdots \forall x_n \varphi$ where φ is quantifier-free. An *existential* formula is one of the form $\exists x_1 \cdots \exists x_n \varphi$ where, again, φ is quantifier-free.

4.12 Preservation under Submodels. The formula ψ is *preserved under submodels* if for all $\mathcal{B}$, $\mathcal{A} \subset \mathcal{B}$ and $a_1, \ldots, a_n \in A$: if $\mathcal{B} \models \psi[a_1, \ldots, a_n]$, then $\mathcal{A} \models \psi[a_1, \ldots, a_n]$.

4.13 Lemma. *Every universal formula is preserved under submodels.*

Proof. This is Exercise 111. ⊣

Part 2 of the next result says that the converse of this is almost true.

4.14 Proposition.

1. *Every model of the universal consequences of a theory can be extended to a model of that theory,*
2. *every formula that is preserved under submodels has a logical equivalent that is universal.*

Proof. 1. Suppose that the model $\mathcal{A}$ satisfies every universal consequence of the theory Γ. By Corollary 4.9, a model $\mathcal{B} \supset \mathcal{A}$ of Γ is (up to isomorphism) the same as a model of $DIAG(\mathcal{A}) \cup \Gamma$. Thus, it suffices to show that $DIAG(\mathcal{A}) \cup \Gamma$ has a model. For this, apply Compactness. If $DIAG(\mathcal{A}) \cup \Gamma$ has no model, then a finite $\Delta \subset DIAG(\mathcal{A})$ exists such that $\Delta \cup \Gamma$ has no model. Suppose that $\delta(a_1, \ldots, a_n)$ is the conjunction of all sentences in Δ, where δ is a formula in the vocabulary of Γ and $a_1, \ldots, a_n \in A$. We have that $\Gamma \models \neg\delta(a_1, \ldots, a_n)$, and hence, by Exercise 7 (page 7),

$$\Gamma \models \forall x_1 \cdots \forall x_n \neg\delta(x_1, \ldots, x_n).$$

Thus, $\forall x_1 \cdots \forall x_n \neg\delta(x_1, \ldots, x_n)$ is a universal consequence of Γ and, therefore, true in $\mathcal{A}$. In particular, $(\mathcal{A}, a)_{a\in A} \models \neg\delta(a_1, \ldots, a_n)$; a contradiction.

2. Assume that the formula $\psi = \psi(x_1, \dots, x_n)$ is preserved under submodels. Let Π be the set of universal formulas $\varphi = \varphi(x_1, \dots, x_n)$ that are a logical consequence of ψ. First, $\Pi \models \psi$. Indeed, suppose that every formula in Π is satisfied by $a_1, \dots, a_n$ in $\mathcal{A}$. By 1, $(\mathcal{A}, a_1, \dots, a_n)$ extends to a model of $\psi(a_1, \dots, a_n)$. Since ψ is preserved under submodels, we have that $\mathcal{A} \models \psi[a_1, \dots, a_n]$.

Next, by Compactness, a finite $\Pi' \subset \Pi$ exists such that $\Pi' \models \psi$. Thus, the conjunction $\bigwedge \Pi'$ is an equivalent of ψ. Finally, $\bigwedge \Pi'$ has a universal equivalent (simply move out universal quantifiers) and this is the required sentence. $\dashv$

Proposition 4.14 is an extremely simple example of a so-called *preservation result*, that is: a theorem characterizing the sentences preserved under some algebraic transformation or relation. Such results exist for sentences preserved under limits of chains, homomorphic images (cf. Theorem 4.55), cartesian products, etc. For one more simple example, see Exercise 116.

Exercises

108 Prove Lemma 4.8.
Hint. Use Lemma 2.15.

109 Suppose that $\mathcal{A}$ is an L-model. Show that the following sets have the same L_A-models:

1. $DIAG(\mathcal{A})$,
2. the set of all quantifier-free L_A-sentences true in $(\mathcal{A}, a)_{a \in A}$ (this contains $DIAG(\mathcal{A})$ as a proper subset),
3. the set of all L_A-sentences true in $(\mathcal{A}, a)_{a \in A}$ that have one of the following five forms, where $a, a', a_1, \dots, a_n$ are new constants from A, and $\mathbf{c}$, $\mathbf{f}$ and $\mathbf{r}$ are (respectively: constant, function and relation) symbols from L:
   a. $\neg a = a'$,
   b. $a = \mathbf{c}$,
   c. $a = \mathbf{f}(a_1, \dots, a_n)$,
   d. $\mathbf{r}(a_1, \dots, a_n)$,
   e. $\neg \mathbf{r}(a_1, \dots, a_n)$

   (this is a proper subset of $DIAG(\mathcal{A})$).

Hint. Show: if $h : A \to B$ is such that $(\mathcal{B}, h(a))_{a \in A}$ is a model of 3a–e, then h embeds $\mathcal{A}$ in $\mathcal{B}$.

110 Modify the notion of diagram such that the following equivalence holds: $(\mathcal{B}, h(a))_{a \in A}$ is a model of this modified diagram of $\mathcal{A}$ iff h is a homomorphism from $\mathcal{A}$ into $\mathcal{B}$.

111 Prove Lemma 4.13.

112 Show that the model $(\mathbb{N}, <, +, \times, 0, 1, 2, \ldots)$ has a countable proper elementary extension. Show that every such *non-standard model* of arithmetic model has order type $\omega + \zeta \cdot \eta$.
Hint. For the first part, choose a fresh constant symbol $\mathbf{c}$ and apply Compactness to the set of sentences

$$Th(\mathbb{N}, <, +, \times, 0, 1, 2, \ldots) \cup \{\mathbf{c} \neq 0, \mathbf{c} \neq 1, \mathbf{c} \neq 2, \ldots\}.$$

(Compare the proof of Theorem 4.10.) For the second part, identify elements of such a non-standard model if the interval determined by them is finite. Then (the interpretations of the constant symbols) $0, 1, 2, \ldots$ form one equivalence class of type ω, the remaining ones all have type ζ; they are themselves ordered in a type that is dense and has no endpoints.

The next two exercises are classical applications of Compactness.

113 A graph (see page 38) is *k-colorable* if you can color its elements with colors $1, \ldots, k$ in such a way that connected elements are colored differently. Show that if every finite subgraph of a graph is k-colorable, then so is the graph itself.
Hint. Apply Compactness to the set that is the union of the diagram of the graph and a set of sentences expressing that certain relations form a k-coloring.

114 Show that every partial ordering extends to a linear ordering. (That is, if $\preceq$ partially orders a set A, then a linear ordering $\leq$ of A exists such that for $a, b \in A$, $a \preceq b$ implies $a \leq b$.)

The next exercises describe the relationship between Horn sentences and submodels of cartesian products.

115 The *cartesian product* of models $\mathcal{A}_i$ $(i \in I)$ is the model $\mathcal{A} = \prod_{i \in I} \mathcal{A}_i$, where $A = \prod_{i \in I} A_i$ (see Section B.2), an individual constant $\mathbf{c}$ is interpreted as the function $\mathbf{c}^{\mathcal{A}}$ defined by $\mathbf{c}^{\mathcal{A}}(i) = \mathbf{c}^{\mathcal{A}_i}$, for an n-ary function symbol $\mathbf{f}$, its interpretation is the n-ary function $\mathbf{f}^{\mathcal{A}}$ over A defined by

$$\mathbf{f}^{\mathcal{A}}(a_1, \ldots, a_n)(i) = \mathbf{f}^{\mathcal{A}_i}(a_1(i), \ldots, a_n(i)),$$

and for every n-ary relation symbol $\mathbf{r}$, its interpretation $\mathbf{r}^{\mathcal{A}}$ is defined by

$$\mathbf{r}^{\mathcal{A}}(a_1, \ldots, a_n) :\equiv \forall i \in I(\mathbf{r}^{\mathcal{A}_i}(a_1(i), \ldots, a_n(i))).$$

Show that

1. if $t = t(x_1, \ldots, x_m)$ is a term and $a_1, \ldots, a_m \in A$, then for all $i \in I$,
$$t^{\mathcal{A}}[a_1, \ldots, a_m](i) = t^{\mathcal{A}_i}[a_1(i), \ldots, a_m(i)],$$
2. if $\varphi = \varphi(x_1, \ldots, x_m)$ is an atomic formula and $a_1, \ldots, a_m \in A$, then
$$\mathcal{A} \models \varphi[a_1, \ldots, a_m]$$
iff for all $i \in I$, $\mathcal{A}_i \models \varphi[a_1(i), \ldots, a_m(i)]$.

116 In the logic programming literature, a *Horn sentence* is a universal quantification of a disjunction of formulas that are atomic or negated atomic, but among which at most one formula is (unnegated) atomic. If there is an unnegated disjunct present, such a sentence can be written as $\forall(\bigwedge \Psi \to \psi)$, where ψ and the elements of Ψ are atoms.

Show that Horn sentences are preserved under cartesian products.

Since Horn sentences are universal, they are also preserved under submodels. Show that

1. every model of the Horn consequences of a theory can be embedded in a product of models of that theory,
2. every sentence that is preserved under submodels of products has a Horn equivalent.

Sketch. 1. Suppose that $\mathcal{B}$ satisfies all Horn consequences of T. Write $DIAG(\mathcal{B}) = P \cup N$, where P contains the atoms and N the negations of atoms from $DIAG(\mathcal{B})$.
Claim. For every $L \in N$ there is a model $\mathcal{B}_L$ of $T \cup P \cup \{L\}$.
Proof. Suppose that $L = \neg\varphi$. If $T \cup P \cup \{L\}$ has no model, then, by Compactness, there exists a finite $Q \subset P$ such that $T \models \bigwedge Q \to \varphi$. Thus,

$$T \models \forall x \cdots (\bigwedge Q' \to \varphi'),$$

where Q' and φ' are obtained form Q and φ by replacing constants from A by new variables $x, \ldots$ Thus, $\forall x \cdots (\bigwedge Q' \to \varphi')$ is a Horn consequence of T and consequently holds in $\mathcal{B}$. Contradiction.

Now, check that the product model $\prod_L \mathcal{B}_L$ satisfies $DIAG(\mathcal{B})$.

4.3 Ultraproducts

The ultrafilters of the next definition can be used to construct a curious species of models: *ultraproducts*. Roughly, these are quotients of cartesian products of models (see Exercise 115), where the congruence is induced by an ultrafilter over the index set.

4.15 Fip, Filters, Ultrafilters. A collection F of subsets of a set I has the *finite intersection property* (*fip*) if no intersection of finitely many elements of F is empty.

F is a *filter* if

1. $I \in F$,
2. $X \in F \wedge X \subset Y \subset I \Rightarrow Y \in F$, and
3. $X, Y \in F \Rightarrow X \cap Y \in F$.

A filter F over I is an *ultrafilter* if for all $X \subset I$, $X \in F$ iff $I - X \notin F$.

For the fact that every fip collection can be extended to an ultrafilter, see Exercise 122.

By convention, I is the intersection of the empty subcollection of F. Thus, for a fip collection to exist, I should be non-empty. Note that a fip collection can very well have an empty intersection itself. An example is the collection of all sets $I - \{i\}$ ($i \in I$) where I is an infinite set.

You can think of the sets in a filter over I as in some sense "big". An ultrafilter partitions the subsets of I into "big" and "small" sets.

4.16 Reduced Products, Ultraproducts, and Ultrapowers. Suppose that $\{\mathcal{A}_i \mid i \in I\}$ is an indexed set of L-models and F a collection of subsets of I that is a filter. The following defines an L-model $\mathcal{A} = \prod_F \mathcal{A}_i$, the *reduced product*, of the models $\mathcal{A}_i$ modulo F.

Define the relation $\sim$ on the cartesian product $\prod_{i \in I} A_i$ by

$$h \sim j \;:\equiv\; \{i \in I \mid h(i) = j(i)\} \in F$$

(h coincides with j on a "big" set). This relation is an equivalence: see Exercise 123. For the universe A of the model to be constructed, take the quotient $\prod_{i \in I} A_i / \sim$, that is: the collection of equivalence classes $|h| := \{j \mid j \sim h\}$.

To describe the L-structure over A, it is convenient to employ λ-*notation.* If, for every $i \in I$, $F(i)$ is the description of an object, then $\lambda i.F(i)$ is a notation for the function F that assigns to every i the object $F(i)$.

Now, let $\mathbf{c}$ be an individual constant from L that is interpreted in $\mathcal{A}_i$ as $\mathbf{c}^{\mathcal{A}_i}$. Its interpretation $\mathbf{c}^{\mathcal{A}}$ in $\mathcal{A}$ is defined as

$$\mathbf{c}^{\mathcal{A}} := |\lambda i.\mathbf{c}^{\mathcal{A}_i}|.$$

Next, let $\mathbf{f}$ be an n-ary function symbol from L that is interpreted in $\mathcal{A}_i$ as $\mathbf{f}^{\mathcal{A}_i}$. Its interpretation $\mathbf{f}^{\mathcal{A}}$ over $\mathcal{A}$ is defined by

$$\mathbf{f}^{\mathcal{A}}(|h_1|, \ldots, |h_n|) := |\lambda i.\mathbf{f}^{\mathcal{A}_i}(h_1(i), \ldots, h_n(i))|.$$

By Exercise 124, this definition does not depend on the representatives $h_1, \ldots, h_n$ chosen in the equivalence classes $|h_1|, \ldots, |h_n|$.

Finally, let $\mathbf{r}$ be an n-ary function symbol from L that is interpreted in $\mathcal{A}_i$ as $\mathbf{r}^{\mathcal{A}_i}$. Its interpretation $\mathbf{r}^{\mathcal{A}}$ over $\mathcal{A}$ is defined by

$$\mathbf{r}^{\mathcal{A}}(|h_1|, \ldots, |h_n|) \;:\equiv\; \{i \in I \mid \mathbf{r}^{\mathcal{A}_i}(h_1(i), \ldots, h_n(i))\} \in F.$$

Again by Exercise 124, this definition does not depend on the representatives $h_1, \ldots, h_n$ chosen in the equivalence classes $|h_1|, \ldots, |h_n|$.

This completes the definition of the reduced product $\prod_F \mathcal{A}_i$. If F happens to be an ultrafilter, the reduced product is called an *ultraproduct.* If all factor models $\mathcal{A}_i$ are the same, a reduced (ultra) product is called a reduced (respectively, ultra) *power.*

The following result describes the fundamental relationship that exists between the evaluation of terms and formulas in an ultraproduct and its factors.

4.17 Fundamental Theorem. *Suppose that $\mathcal{A} = \prod_F \mathcal{A}_i$ is an ultraproduct and $|h_1|, \ldots, |h_n|$ a sequence of elements from A.*

1. *For every term $t = t(x_1, \ldots, x_n)$,*
$$t^{\mathcal{A}}[|h_1|, \ldots, |h_n|] = |\lambda i.t^{\mathcal{A}_i}[h_1(i), \ldots, h_n(i)]|,$$
2. *for every formula $\varphi = \varphi(x_1, \ldots, x_n)$,*
$$\mathcal{A} \models \varphi[|h_1|, \ldots, |h_n|] \Leftrightarrow \{i \in I \mid \mathcal{A}_i \models \varphi[h_1(i), \ldots, h_n(i)]\} \in F,$$
3. *for every sentence φ,*
$$\mathcal{A} \models \varphi \Leftrightarrow \{i \in I \mid \mathcal{A}_i \models \varphi\} \in F.$$

Proof. Part 1 is proved by a straightforward term induction and does not need F to be ultra. Part 2 is proved by induction on φ. The atomic case follows from 1 and the way interpretations for relation symbols have been defined; the cases for the connectives follow from the filter properties (in particular, the case for negation needs that F is ultra). Here follows the case for the existential quantifier.
(i) Assume that $\mathcal{A} \models \exists x_0 \varphi[|h_1|, \ldots, |h_n|]$. For instance,
$$\mathcal{A} \models \varphi[|h_0|, |h_1|, \ldots, |h_n|].$$
By induction hypothesis,
$$\{i \in I \mid \mathcal{A}_i \models \varphi[h_0(i), h_1(i), \ldots, h_n(i)]\} \in F.$$
Note that
$$\{i \in I \mid \mathcal{A}_i \models \varphi[h_0(i), h_1(i), \ldots, h_n(i)]\}$$
is a subset of $\{i \in I \mid \mathcal{A}_i \models \exists x_0 \varphi[h_1(i), \ldots, h_n(i)]\}$. Thus,
$$\{i \in I \mid \mathcal{A}_i \models \exists x_0 \varphi[h_1(i), \ldots, h_n(i)]\} \in F.$$
(ii) Assume that $X := \{i \in I \mid \mathcal{A}_i \models \exists x_0 \varphi[h_1(i), \ldots, h_n(i)]\} \in F$. By the Axiom of Choice, choose the function h_0 in $\prod_{i \in I} A_i$ in such a way that if $i \in X$, then
$$\mathcal{A}_i \models \varphi[h_0(i), h_1(i), \ldots, h_n(i)]\}.$$
Thus,
$$X \subset \{i \in I \mid \mathcal{A}_i \models \varphi[h_0(i), h_1(i), \ldots, h_n(i)]\}.$$
Therefore,
$$\{i \in I \mid \mathcal{A}_i \models \varphi[h_0(i), h_1(i), \ldots, h_n(i)]\} \in F.$$
By induction hypothesis, $\mathcal{A} \models \varphi[|h_0|, |h_1|, \ldots, |h_n|]$. Thus,
$$\mathcal{A} \models \exists x_0 \varphi[|h_1|, \ldots, |h_n|].$$
See Exercise 129 for further details. $\dashv$

As a corollary of this result, here follows an *ultra short* proof for the Compactness Theorem. It shows that you can construct a model for a collection of sentences *directly*, by taking a suitable ultraproduct of models of *finite subsets* of the collection.

Proof of Theorem 4.1. Suppose that Σ is a set of sentences such that every finite $\Delta \subset \Sigma$ has a model $\mathcal{A}_\Delta$. Let I be the set of these finite subsets Δ. What we look for is an ultrafilter F over I such that $\prod_F \mathcal{A}_\Delta \models \Sigma$. Now, what is needed of F? Let $\varphi \in \Sigma$ be arbitrary. By Theorem 4.17.3, in order that $\prod_F \mathcal{A}_\Delta \models \varphi$ (where F is still hypothetical), it suffices to have that $\{\Delta \mid \mathcal{A}_\Delta \models \varphi\} \in F$. Note that for every $\Delta \in I$, $\mathcal{A}_\Delta \models \Delta$. Thus, $\varphi \in \Delta$ implies $\mathcal{A}_\Delta \models \varphi$, and therefore $\{\Delta \mid \varphi \in \Delta\} \subset \{\Delta \mid \mathcal{A}_\Delta \models \varphi\}$.

Conclusion: it just suffices to have all sets $\widehat{\varphi} := \{\Delta \mid \varphi \in \Delta\}$ $(\varphi \in \Sigma)$ in the ultrafilter. And that this is possible follows from the:

Claim. The collection $\{\widehat{\varphi} \mid \varphi \in \Sigma\}$ has the fip.
Proof. The (extremely simple) verification should not take you longer than, say, 15 seconds. However, if this baffles you, you can find the solution in Exercise 130. $\dashv$

Thus, by Exercise 122, there is indeed an ultrafilter F that contains every $\widehat{\varphi}$, and the proof is complete. $\dashv$

Apologies for the fact that the essential ingredient for this proof, Exercise 122, according to its second *hint*, should be solved by means of Compactness!

4.18 Canonical embedding. Suppose that $\prod_F \mathcal{A}$ is the ultrapower of the model $\mathcal{A}$ that is determined by an ultrafilter F over the set I. The *canonical embedding* of $\mathcal{A}$ into $\prod_F \mathcal{A}$ is the function j defined by $j(a) := |\lambda i.a|$. It maps an element a to the equivalence class of the constant function $i \mapsto a$.

4.19 Lemma. *Every canonical embedding is elementary.*

Proof. By Theorem 4.17, $\prod_F \mathcal{A} \models \varphi[j(a_1), \ldots, j(a_n)]$ holds iff

$$\{i \mid \mathcal{A} \models \varphi[a_1, \ldots, a_n]\} \in F;$$

the latter condition amounts to $\mathcal{A} \models \varphi[a_1, \ldots, a_n]$. $\dashv$

Exercises

117 Suppose that I is a set. Show: if $X \subset I$, then $\{Y \subset I \mid X \subset Y\}$ is a filter. Show that this is an ultrafilter iff X is a singleton.

118 Assume that F is a fip collection of subsets of I. Show that the collection of subsets X of I such that for some finite $Y_1, \ldots, Y_n \in F$,

$$Y_1 \cap \cdots \cap Y_n \subset X,$$

is a filter.

119 Suppose that F is an ultrafilter over I. Show that if $X_1 \cup \cdots \cup X_n \in F$, then for some i, $X_i \in F$.

120 Show that the collection of sets that are closed and unbounded in some uncountable regular initial number is a filter.

121 Show that ultrafilters are the same as maximal fip collections.

122 Show that every fip collection F over a non-empty set I can be extended to an ultrafilter.
Hints.
1. The result of Exercise 121 suggests using Zorn's Lemma.
2. Here follows a sketch that employs Compactness. Note that if $\bigcap F$ is non-empty, then (by Exercise 117) it is very easy to find the required ultrafilter: pick any $i \in \bigcap F$ and consider $\{X \subset I \mid i \in X\}$. By the way, an ultrafilter of this form is called *principal.* But, as was illustrated above, you might easily have that $\bigcap F = \emptyset$. The trick now is that *from a certain perspective* $\bigcap F$ still happens to be non-empty. Consider the model $\mathcal{A} := (\mathcal{P}(I), \subset, \cap, ', X)_{X \subset I}$, where $\mathcal{P}(I) = \{X \mid X \subset I\}$ is the collection of all subsets of I and $\subset, \cap, '$ denote set inclusion, intersection and complementation with respect to I.

Show that $\mathcal{A}$ has an elementary extension $\mathcal{B}$ that has an element $b \in B$ such that $\mathcal{B} \models b \neq \emptyset \wedge \forall y(b \subset y \vee b \subset y')$ (so, from the perspective of $\mathcal{B}$, b behaves like a one-element set) and such that for all $X \in F$, $\mathcal{B} \models b \subset X$ (from the perspective of $\mathcal{B}$, the one element in b is in every $X \in F$).

Although of course b is not really a one-element set, it still yields the required ultrafilter: show that the collection of $X \subset I$ for which $\mathcal{B} \models b \subset X$ satisfies all requirements.

123 Show that the relation $\sim$ defined in Definition 4.16 is indeed an equivalence relation (i.e., reflexive, symmetric, and transitive).
Hint. For instance, transitivity: assume that $h \sim j$ and $j \sim k$. That is, $\{i \in I \mid h(i) = j(i)\}, \{i \in I \mid j(i) = k(i)\} \in F$. Note that

$$\{i \in I \mid h(i) = j(i)\} \cap \{i \in I \mid j(i) = k(i)\} \subset \{i \in I \mid h(i) = k(i)\}.$$

Thus, by the filter properties, $\{i \in I \mid h(i) = k(i)\} \in F$; i.e., $h \sim k$.

124 Show that, in Definition 4.16, the definition of the interpretation of function and relation symbols in the reduced product does not depend on the representatives chosen in the equivalence classes that occur as arguments.
Hint. Compare the example given in Exercise 123.

125 Suppose that F is the principal ultrafilter over I that is determined by $i_0 \in I$; that is: $F = \{X \subset I \mid i_0 \in I\}$. Show that $\prod_F \mathcal{A}_i \cong \mathcal{A}_{i_0}$.

126 Suppose that F is the filter $\{I\}$ over I. Show that $\prod_F \mathcal{A}_i \cong \prod_{i \in I} \mathcal{A}_i$. (Cf. Exercise 115.)

127 Suppose that F is a filter over I. Show that $\prod_F \mathcal{A}_i$ is a homomorphic image of $\prod_{i \in I} \mathcal{A}_i$.

128 Show that if every model $\mathcal{A}_i$ has at most n elements, then every ultraproduct $\prod_F \mathcal{A}_i$ has at most n elements. Give an example of an ultraproduct of finite models that is infinite.

129 Complete the proof of Theorem 4.17.

130 Show that the collection $\{\widehat{\varphi} \mid \varphi \in \Sigma\}$ that occurs in the ultraproduct proof of the Compactness Theorem is fip.
Hint. If $\varphi_1, \ldots, \varphi_n \in \Sigma$, then $\{\varphi_1, \ldots, \varphi_n\} \in \widehat{\varphi_1} \cap \cdots \cap \widehat{\varphi_n}$.

131 Show that if $\mathcal{A}$ is finite and F ultra, then $\prod_F \mathcal{A} \cong \mathcal{A}$.

132 Consider the linear ordering $\omega = (\mathbb{N}, <)$. Let F be an ultrafilter over $\mathbb{N}$ that contains the fip collection of all sets $\{n, n+1, n+2, \ldots\}$. Build the ultrapower $\mathcal{A} := \prod_F \omega$. $\mathcal{A}$ is a linear ordering. Let j be the canonical embedding from ω into $\mathcal{A}$. Show that for every n, $j(n)$ is the n-th element of $\mathcal{A}$.

Let $h := \lambda n.n$ be the identity function on $\mathbb{N}$. Show that for every $n \in \mathbb{N}$, $\mathcal{A} \models j(n) < |h|$. (Thus, $\mathcal{A}$ does not have order type ω.)

Investigate the position of other equivalence classes of functions in $\mathcal{A}$.

Show that the order type of $\mathcal{A}$ has the form $\omega + \zeta \cdot \delta$, where δ is a dense type without endpoints.

133 Show that an ultrapower $\prod_F \omega$ is isomorphic to its base model ω iff the ultrafilter F is *ω-complete*, that is: for all $X_0, X_1, X_2, \ldots \in F$, $\bigcap_n X_n \in F$.
Hint. (Only if.) If $X_0, X_1, X_2, \ldots \in F$ but $\bigcap_n X_n \notin F$, let $h : I \to \mathbb{N}$ be such that if $i \in X_0 \cap \cdots X_{n-1}$ and $i \notin X_n$, then $h(i) = n$. Show that $|h|$ cannot be the n-th element in the ordering of the ultrapower for any $n \in \mathbb{N}$.

(If.) Show that the ordering of the ultrapower is well-ordered (has no infinite descending sequence $|h_0| > |h_1| > |h_2| > \cdots$).

134 Assume that $\mathcal{A} \equiv \mathcal{B}$. Show that $\mathcal{A}$ can be elementarily embedded into an ultrapower of $\mathcal{B}$.
Hint. For every finite $\Delta \subset ELDIAG(\mathcal{A})$, there is a simple expansion $\mathcal{B}_\Delta$ of $\mathcal{B}$ that is a model of Δ. Thus, $ELDIAG(\mathcal{A})$ has a model of the form $\prod_F \mathcal{B}_\Delta$.

135 Suppose that the vocabulary of $\mathcal{A}$ does not contain individual constants or function symbols. Show that $\mathcal{A}$ can be embedded into an ultraproduct of its finite submodels.

136 If Σ is a set of sentences, then $Mod(\Sigma) := \{\mathcal{A} \mid \mathcal{A} \models \Sigma\}$. Let K be class of models. Show the following:

1. There exists a set of sentences Σ such that $K = Mod(\Sigma)$ iff K is closed under ultraproducts and equivalence, i.e.: (i) every ultraproduct of models from K is in K, and (ii) every equivalent of a model in K is in K.

2. There exists a sentence σ such that $K = Mod(\{\sigma\})$ iff both K and its complement are closed under ultraproducts and equivalence.

Hint. By Compactness, 2 follows from 1.

137 Suppose that $\mathcal{A} \prec \mathcal{B}$. Show that there is an elementary embedding h from $\mathcal{B}$ into an ultrapower of $\mathcal{A}$, such that the restriction $h \mid A$ of this embedding to A coincides with the canonical embedding from $\mathcal{A}$ into the ultrapower.

138 Suppose that $\mathcal{A}_0 \subset \mathcal{A}_1 \subset \mathcal{A}_2 \subset \cdots$ is a strictly ascending chain of models with limit $\mathcal{A}$. Let F be a filter over $\mathbb{N}$ that contains all sets $\{n, n+1, n+2, \ldots\}$. Show that $\mathcal{A}$ can be embedded in $\prod_F \mathcal{A}_i$.

By Theorem 4.17, all first-order sentences are preserved by ultraproducts. The next exercise states that Σ^1_1-sentences are preserved as well.

139 Suppose that the Σ^1_1-sentence Φ (Definition 3.33 page 36) is satisfied by every factor $\mathcal{A}_i$ of the ultraproduct $\prod_F \mathcal{A}_i$. Show that $\prod_F \mathcal{A}_i \models \Phi$.

4.4 Omitting Types

The proof of the Compactness Theorem above is a rather crude application of the Henkin construction. A more refined argument produces the Omitting Types Theorem.

From now on *all vocabularies will be countable.*

4.20 Types. A *k-type* is a set $\tau = \tau(\vec{x})$ of formulas in the free variables $\vec{x} = (x_1, \ldots, x_k)$. A 1-type often is called simply a *type.*

If some sequence $\vec{a}$ from A satisfies all formulas of the type τ in $\mathcal{A}$, then $\mathcal{A}$ is said to *realize* τ. $\mathcal{A}$ *omits* τ if it is not realized in $\mathcal{A}$.

A type is a type *of* the theory Γ if it is realized in a model of Γ. It is a type *of* $\mathcal{A}$ if it is a type of $Th(\mathcal{A})$.

A type τ of Γ is *principal* if there is a formula $\varphi = \varphi(\vec{x})$ — a *generator* of τ — satisfiable in a model of Γ and such that $\Gamma \models (\varphi \rightarrow \bigwedge \tau)$ (i.e.: $\Gamma \models \varphi \rightarrow \psi$ for all $\psi \in \tau$). It is a *principal* type of $\mathcal{A}$ if it is principal with respect to $Th(\mathcal{A})$.

(It is helpful to realize that there exists a parallel —that can be made precise— between the notions of (*principal*) *type* and (*principal*) *filter*.)

Note that a model does not need to realize each one of its types. Example: $\mathcal{A} = (\mathbb{N}, 0, 1, 2, \ldots)$, $\tau = \{x \neq 0, x \neq 1, x \neq 2, \ldots\}$. However, a model does always realize its principal types. (If φ is a generator of some type of $\mathcal{A}$, then $\mathcal{A} \models \exists \vec{x} \varphi$.)

In order that the reader becomes more familiar with the notion of type, he is urged to check the details of the following lemma.

4.21 Lemma. *For a type $\tau = \tau(\vec{x})$ in the vocabulary of a model $\mathcal{A}$, the following are equivalent:*

1. *τ is a type of $\mathcal{A}$; i.e.: τ is realized in some $\mathcal{B} \equiv \mathcal{A}$,*
2. *τ is finitely satisfiable in $\mathcal{A}$,*
3. *τ is a type of $ELDIAG(\mathcal{A})$; i.e.: τ is realized in some $\mathcal{B} \succ \mathcal{A}$.*

Proof. $1 \Rightarrow 2$. Suppose that (1) $\vec{b}$ satisfies τ in $\mathcal{B} \equiv \mathcal{A}$. If $\tau' \subset \tau$ is finite, then $\mathcal{B} \models \exists\vec{x} \bigwedge \tau'$. Since $\mathcal{B} \equiv \mathcal{A}$, we have that $\mathcal{A} \models \exists\vec{x} \bigwedge \tau'$ as well. Thus, 2 holds.
$2 \Rightarrow 3$. Suppose that (2) τ is finitely satisfiable in $\mathcal{A}$. Then $\tau \cup ELDIAG(\mathcal{A})$ is finitely satisfiable. By Compactness, τ is realized in some $\mathcal{B} \succ \mathcal{A}$.
$3 \Rightarrow 1$. Trivial. $\dashv$

4.22 Omitting Types Theorem. *Suppose that Γ is a satisfiable set of sentences and T is a countable set of types that are non-principal with respect to Γ. Then Γ has a (countable) model omitting every type from T.*

Proof. Let L be the countable vocabulary involved. Choose a countable set C of fresh individual constants. Put $L' := L \cup C$. Construct a maximal satisfiable set $\Gamma^\star \supset \Gamma$ with the Henkin property such that if $\tau = \tau(\vec{x}) \in T$ and $\vec{\mathbf{c}}$ is a sequence from C that has the same length as $\vec{x}$, then $\psi \in \tau$ exists such that $\neg\psi(\vec{\mathbf{c}}) \in \Gamma^\star$. It follows from Exercise 100 that the canonical model for $\Gamma^\star$ omits all types from T.

$\Gamma^\star$ is the limit of a sequence $\Gamma_0 = \Gamma \subset \Gamma_1 \subset \Gamma_2 \subset \cdots$ that is constructed as follows. Fix enumerations of (i) all L'-sentences and (ii) of all pairs $(\tau, \vec{\mathbf{c}})$ where $\tau = \tau(\vec{x}) \in T$ and $\vec{\mathbf{c}}$ is a sequence of appropriate length from C. Γ_{n+1} is obtained by adding one, two or three sentences to Γ_n as follows.

1. Add the n-th L'-sentence of the first enumeration if this does not result in unsatisfiability. (This produces *maximal* satisfiability of $\Gamma^\star$.)
2. If the n-th L'-sentence is added at step 1 and has the form $\exists x\varphi(x)$, choose a $\mathbf{c} \in C$ that does not occur in φ or in a sentence from Γ_n (such a $\mathbf{c}$ exists, since $\Gamma_n - \Gamma$ is finite) and add $\varphi(\mathbf{c})$. (This ascertains that $\Gamma^\star$ has the Henkin property.)
3. If $(\tau, \vec{\mathbf{c}})$ is the n-th pair from the second enumeration, add a sentence $\neg\psi(\vec{\mathbf{c}})$, where $\psi \in \tau$ is such that this addition preserves satisfiability. (This results in a canonical model omitting all types from T.)

Since in 2., $\mathbf{c}$ is fresh and $\exists x\varphi \in \Gamma_n$, addition of $\varphi(\mathbf{c})$ does not spoil satisfiability.

Finally, it must be shown that, in (3), such a $\psi \in \tau$ always exists. Let φ be the conjunction of all sentences added to Γ up to that point of the construction. Write $\varphi = \delta(\vec{\mathbf{c}}, \vec{\mathbf{a}})$, where $\vec{\mathbf{a}}$ is the sequence of constant symbols from C in φ that are not in $\vec{\mathbf{c}}$. (Not every constant from $\vec{\mathbf{c}}$ need actually appear in φ.) If the addition of $\neg\psi(\vec{\mathbf{c}})$ results in an unsatisfiable set, then

$$\Gamma \models \delta(\vec{\mathbf{c}}, \vec{\mathbf{a}}) \rightarrow \psi(\vec{\mathbf{c}}).$$

Thus (since the constants from $\vec{\mathbf{a}}$ do not occur in Γ or in $\psi(\vec{\mathbf{c}})$, by Exercise 7 (page 7)

$$\Gamma \models \exists \vec{y}\delta(\vec{\mathbf{c}}, \vec{y}) \rightarrow \psi(\vec{\mathbf{c}})$$

and so (again by Exercise 7)

$$\Gamma \models \forall \vec{x}[\exists \vec{y}\delta(\vec{x}, \vec{y}) \rightarrow \psi(\vec{x})].$$

If this holds for *every* $\psi \in \tau$, then clearly $\exists \vec{y}\delta(\vec{x}, \vec{y})$ is a generator of τ, *unless* it is not satisfiable in a model of Γ. However, by construction, φ is satisfiable in a model of Γ, and hence $\exists \vec{y}\delta(\vec{x}, \vec{y})$ is satisfiable in a model of Γ as well. ⊣

In the above proof, there is nothing that is really typical for first-order logic. The same construction proves the result for countable fragments of infinitary logic.

Suppose that the binary relation symbol $<$ is in L. The *Collection Principle* is the set of all formulas of the form

$$\forall x < a\, \exists y\, \varphi \;\rightarrow\; \exists b\, \forall x < a\, \exists y < b\, \varphi,$$

where φ is a formula not containing b freely. For example, the standard model of arithmetic, ordered in type ω, is a model of the Collection Principle. More generally, every ordered model that has a regular order type satisfies Collection.

$\mathcal{B}$ is an *end extension* of $\mathcal{A}$ if for all $a \in A$ it holds that

$$b \in B \wedge b < a \Rightarrow b \in A$$

that is: no element of A gets a new $<$-predecessor in B.

4.23 Proposition. *Every countable linearly ordered model of Collection has a proper elementary end extension.*

Proof. Let $\mathcal{A}$ be a countable linearly ordered model of the Collection Principle. Let E be the elementary diagram of $\mathcal{A}$ and $\mathbf{c}$ a fresh individual constant. Put $O := \{a < \mathbf{c} \mid a \in A\}$. What we want is a model of $E \cup O$ that omits every type

$$\tau_a := \{x < a\} \cup \{x \neq b \mid b < a\}.$$

By the Omitting Types Theorem, it suffices to show that these types are non-principal with respect to $E \cup O$. Assume that $\varphi(x, \mathbf{c})$ generates τ_a. Then, for $b < a$:

$$E \cup O \models \varphi(x, \mathbf{c}) \rightarrow b \neq x;$$

hence $E \cup O \models \neg\varphi(b, \mathbf{c})$. By Compactness there are finitely many $a_1, \ldots,$ $a_k \in A$ such that $E \models a_1, \ldots, a_k < \mathbf{c} \rightarrow \neg\varphi(b, \mathbf{c})$. Let m be the maximum of $a_1, \ldots, a_k$. Then: $E \models m < \mathbf{c} \rightarrow \neg\varphi(b, \mathbf{c})$, and hence

$$E \models m < y \rightarrow \neg\varphi(b, y).$$

Concluding: in $\mathcal{A}$ it holds that for all $a \in A$: $\forall x < a \exists m \forall y > m \neg\varphi(x, y)$. By Collection, there exists $m' \in A$ such that

$$\forall x < a\, \exists m < m'\, \forall y > m\, \neg\varphi(x, y),$$

and hence: $\forall x < a\, \forall y > m'\, \neg\varphi(x, y)$. But then

$$E \cup O \models \forall x < a\, \neg\varphi(x, \mathbf{c}),$$

i.e.: $E \cup O \models \varphi(x, \mathbf{c}) \to \neg x < a$. However, we also have that $E \cup O \models \varphi(x, \mathbf{c}) \to x < a$. Thus, $\varphi(x, \mathbf{c})$ is not satisfiable in a model of $E \cup O$, a contradiction. $\dashv$

Löwenheim-Skolem theorems are about cardinals of universes. *Two-cardinal theorems* consider the cardinal of the universe together with the cardinal of a designated unary relation.

Here is the simplest of examples. Suppose that $U \in L$ is a unary relation symbol.

4.24 Vaught's Two-cardinal Theorem. *If $\mathcal{A}$ is a model in which*

$$\aleph_0 \leq |U^{\mathcal{A}}| < |A|,$$

then $\mathcal{A}$ has an elementary equivalent $\mathcal{B}$ such that

$$|B| = \aleph_1 \text{ and } |U^{\mathcal{B}}| = \aleph_0.$$

Proof. By the Downward Löwenheim-Skolem theorem, you can cut down the cardinality of $|A|$ to the first cardinal $> |U^{\mathcal{A}}|$, keeping $U^{\mathcal{A}}$ fixed. So, without loss of generality, you may assume that already $|A|$ itself is of this power.

Let $<$ be a new binary relation symbol. Choose a well-ordering $<^{\mathcal{A}}$ of A of initial type. Since successor cardinals are regular (see Section B.5), $(\mathcal{A}, <^{\mathcal{A}})$ satisfies a Collection Principle consisting of all formulas

$$\forall x \in U\, \exists y\, \varphi \;\to\; \exists b\, \forall x \in U\, \exists y < b\, \varphi.$$

(To make things slightly more readable, $x \in U$ is written instead of $U(x)$.)

Let $(\mathcal{B}, <^{\mathcal{B}})$ be a countable equivalent of $(\mathcal{A}, <^{\mathcal{A}})$. The proof method of Proposition 4.23 shows that $\tau = \{U(x)\} \cup \{x \neq b \mid b \in U^{\mathcal{B}}\}$ is a non-principal type of the theory Γ that consists of the elementary diagram of $(\mathcal{B}, <^{\mathcal{B}})$ plus all sentences $b < \mathbf{c}$ ($b \in B$, $\mathbf{c}$ a fresh constant). By the Omitting Types Theorem 4.22, Γ has a countable model $(\mathcal{C}, <^{\mathcal{C}})$ that omits τ. Clearly, $(\mathcal{C}, <^{\mathcal{C}})$ is a countable proper elementary extension of $(\mathcal{B}, <^{\mathcal{B}})$ that has the *same* interpretation of U. Repeating this, we construct an elementary chain of length ω_1 of countable models, while keeping the interpretation of U fixed along the chain. The limit of this chain is the required model.

Here follows the argument that τ is non-principal with respect to Γ. Suppose that $\psi(x, \mathbf{c})$ would be a generator. Then in particular,

$\Gamma \models \neg\psi(b, \mathbf{c})$ for $b \in U^{\mathcal{B}}$. Fix a $b \in U^{\mathcal{B}}$. By Compactness, for some $b_1, \ldots, b_n \in B$ we have that

$$ELDIAG(\mathcal{B}, <^{\mathcal{B}}) \cup \{b_1, \ldots, b_n < \mathbf{c}\} \models \neg\psi(b, \mathbf{c}).$$

Thus, (for $x := b$ and y the maximum of $b_1, \ldots, b_n$) $(\mathcal{B}, <^{\mathcal{B}}) \models \forall x \in U\, \exists y\, \forall z > y\, \neg\psi(x, z)$. By Collection, for some b,

$$(\mathcal{B}, <^{\mathcal{B}}) \models \forall x \in U\, \exists y < b\, \forall z > y\, \neg\psi(x, z),$$

and so $(\mathcal{B}, <^{\mathcal{B}}) \models \forall x \in U\, \forall z > b\, \neg\psi(x, z)$. But then, $\Gamma \models \forall x \in U \neg\psi(x, \mathbf{c})$, i.e., $\Gamma \models \psi(x, \mathbf{c}) \rightarrow x \notin U$. Also, $\Gamma \models \psi(x, \mathbf{c}) \rightarrow x \in U$. Hence, $\psi(x, \mathbf{c})$ cannot be satisfiable in a model of Γ. ⊣

4.25 Prime models. A model is *prime* if it does not realize any of its non-principal types.

4.26 Proposition. *Every two equivalent prime models are partially isomorphic.*

Proof. Suppose that $\mathcal{A}$ and $\mathcal{B}$ are equivalent prime models. The strategy of Sy is to make sure that after his n-th move a position $\{(a_1, b_1), \ldots, (a_n, b_n)\}$ is reached such that $(\mathcal{A}, a_1, \ldots, a_n) \equiv (\mathcal{B}, b_1, \ldots, b_n)$. To see that Sy has an $(n+1)$-st move, assume that Di plays $a_{n+1} \in A$. Consider the $(n+1)$-type

$$\tau := \{\varphi \mid \mathcal{A} \models \varphi[a_1, \ldots, a_{n+1}]\}$$

of $(a_1, \ldots, a_{n+1})$ in $\mathcal{A}$. Let $\varphi = \varphi(x_1, \ldots, x_{n+1})$ generate this type. Note that $\mathcal{A} \models \exists x_{n+1}\varphi[a_1, \ldots, a_n]$. (For, $\mathcal{A} \models \varphi[a_1, \ldots, a_n]$. Otherwise, $\neg\varphi \in \tau$, hence $Th(\mathcal{A}) \models \varphi \rightarrow \neg\varphi$, $Th(\mathcal{A}) \models \neg\varphi$, and φ would not be satisfiable in a model of $Th(\mathcal{A})$.) By induction hypothesis, $\mathcal{B} \models \exists x_{n+1}\varphi[b_1, \ldots, b_n]$. As a counter-move for Sy choose $b_{n+1} \in B$ such that $\mathcal{B} \models \varphi[b_1, \ldots, b_{n+1}]$. ⊣

4.27 Corollary. *Every two countable, equivalent prime models are isomorphic.*

Proof. This is immediate from Proposition 4.26 and Theorem 3.48. ⊣

Recall that a complete theory is a set of sentences that (Definition 3.40, page 41), has but one model up to equivalence; it is $\aleph_0$-*categorical* if, up to isomorphism, it has only one countable model. For an example, see Corollary 3.49, page 45.

4.28 Theorem. *Let Γ be a complete theory in a countable vocabulary. The following are equivalent:*

1. *Γ is $\aleph_0$-categorical,*
2. *all types of Γ are principal,*
3. *for every k, there are only finitely many formulas $\varphi = \varphi(x_1, \ldots, x_k)$ pairwise inequivalent with respect to Γ,*
4. *every model of Γ is prime.*

Proof. 1⇒2. If Γ has a non-principal type, there is one countable model that realizes it and one that omits it; and these models cannot be isomorphic.
2⇒3. Assume that 2 holds, but that 3 doesn't. By Zorn's Lemma, every type of Γ is subtype of a maximal one (cf. Lemma 4.35, page 73).
Claim. Γ has infinitely many maximal k-types.
For suppose not. Let $\varphi_1, \ldots, \varphi_n$ be the generators of these types with $x_1, \ldots, x_k$ free. Then every formula in these free variables has an equivalent that is a disjunction of finitely many φ_i. There are 2^n such formulas. This proves the Claim.

Choose a generator for each of those infinitely many types. The set of negations of these generators is a non-principal type. See Exercise 141.
3⇒2. Let τ be a non-principal type. Without loss of generality it can be assumed that τ is maximal. Pick $\varphi_0 \in \tau$. By assumption, φ_0 is not a generator of τ; therefore $\varphi_1 \in \tau$ exists such that $\varphi_0 \wedge \neg\varphi_1$ is satisfiable in a model of Γ. By maximality, $\varphi_0 \wedge \varphi_1 \in \tau$; but again, this cannot be a generator. Thus, $\varphi_2 \in \tau$ exists such that $\varphi_0 \wedge \varphi_1 \wedge \neg\varphi_2$ is satisfiable in a model of Γ. Repeating this argument produces infinitely many pairwise incomparable formulas $\varphi_0 \wedge \neg\varphi_1$, $\varphi_0 \wedge \varphi_1 \wedge \neg\varphi_2$, $\varphi_0 \wedge \varphi_1 \wedge \varphi_2 \wedge \neg\varphi_3, \ldots$, all satisfiable in a model of Γ.
2⇒4. Trivial.
4⇒1. By Corollary 4.27. ⊣

Fixed point logic. Assume that $\varphi = \varphi(x_1, \ldots, x_k)$ is an $L \cup \{\mathbf{r}\}$-formula that is $\mathbf{r}$-positive (see page 32: every occurrence of $\mathbf{r}$ in φ is in the scope of an even number of negation symbols *and* φ does not contain $\rightarrow$ or $\leftrightarrow$). Let $\mathcal{A}$ be an L-model. Then φ induces a (*elementary*) monotone operator $\Gamma = \Gamma_\varphi : \mathcal{P}(A^k) \rightarrow \mathcal{P}(A^k)$ (which maps k-ary relations over A to k-ary relations) defined by

$$\Gamma(S) := \{(a_1, \ldots, a_k) \in A^k \mid (\mathcal{A}, S) \models \varphi[a_1, \ldots, a_k]\}.$$

(That Γ is monotone, i.e., that $S_1 \subset S_2 \subset A^k$ implies $\Gamma(S_1) \subset \Gamma(S_2)$, is due to φ being $\mathbf{r}$-positive.) Let $\Gamma\!\uparrow = \Gamma_\varphi\!\uparrow$ be the least fixed point of this operator. (See Section B.7.)

Fixed-point logic accommodates a notation for such least fixed points. More precisely, it has the following formula-formation rule:

> if $\varphi = \varphi(x_1, \ldots, x_k, y_1, \ldots, y_m)$ is an $L \cup \{\mathbf{r}\}$-fixed-point formula containing the k-ary relation symbol $\mathbf{r}$ positively with $x_1, \ldots, x_k, y_1, \ldots, y_m$ free, then
>
> $$(z_1, \ldots, z_k) \in \mu\mathbf{r}x_1 \cdots x_k\varphi$$
>
> is an L-fixed-point formula with $z_1, \ldots, z_k, y_1, \ldots, y_m$ free.

The semantics of fixed-point formulas is given by the least fixed point interpretation that stipulates $\vec{c} \in \mu \mathbf{r}\vec{x}\varphi(\vec{x}, \vec{a})$ to be true of $\mathcal{A}$ iff $\vec{c}$ is in the least fixed point of the operator defined by $\varphi(\vec{x}, \vec{a})$.

The fixed-point operator μ can be used to express properties that are not first-order expressible. For instance, $z \in \mu \mathbf{r}x[\forall y(y < x \rightarrow \mathbf{r}(y))]$ defines the set of standard integers in any non-standard model of arithmetic. More generally, it defines the *well-founded part* — the largest well-founded initial — of a model (Exercise 142; see Section B.7). However:

4.29 Proposition. *Over an $\aleph_0$-categorical theory, every fixed-point formula has a first-order equivalent.*

Proof. The *finite* stages of the least fixed point hierarchy corresponding to a definable operator are first-order definable (use the recursive definition). But then, by Theorem 4.28.3, there can only be finitely many of them. Thus, the hierarchy has a finite closure ordinal and the least fixed point is definable. ⊣

This proposition has been used to extend the so-called *0–1-law* from first-order logic to fixed-point logic. This law says that for every purely relational sentence φ of these languages — as well as a couple of others — the fraction of models with universe $\{0, \ldots, n\}$ that satisfy φ either tends to 0 or to 1 when n approaches infinity.

Exercises

140 Suppose that $\mathcal{A}$ is a countable model that is prime and that $\mathcal{B} \equiv \mathcal{A}$. Show that $\mathcal{A}$ can be elementarily embedded into $\mathcal{B}$.
Hint. Fix an enumeration $A = \{a_0, a_1, a_2, \ldots\}$. Find $b_0, b_1, b_2, \ldots \in B$ such that for every n, $(\mathcal{A}, a_0, \ldots, a_{n-1}) \equiv (\mathcal{B}, b_0, \ldots, b_{n-1})$. The correspondence $a_i \mapsto b_i$ is the required embedding.

141 Fill in the details of the proof of Theorem 4.28, 2⇒3. Prove the Claim, and the fact that the negations of all generators form a non-principal type of Γ.
Hint. If $\tau = \tau(x_1, \ldots, x_k)$ is a maximal type of Γ and $\varphi = \varphi(x_1, \ldots, x_k) \notin \tau$, then $\neg\varphi \in \tau$.

142 Show that the fixed point formula $z \in \mu \mathbf{r}x[\forall y(y < x \rightarrow \mathbf{r}(y))]$ defines the *well-founded part* (the largest subset of the universe on which $<$ is well-founded) of any model.

4.5 Saturation

A saturated model realizes many types. More precisely:

4.30 Saturation. A model $\mathcal{A}$ is *saturated* if every simple expansion $(\mathcal{A}, a_1, \ldots, a_n)$ of $\mathcal{A}$ with finitely many elements $a_1, \ldots, a_n \in A$ realizes all its 1-types.

(The concept defined is usually referred to as ω-*saturation*, indicating the limitation to simple expansions with *finitely* ($< \omega$) many elements. But we shall not consider κ-saturation for $\kappa > \omega$.)

The restriction to 1-types in Definition 4.30 can be lifted:

4.31 Lemma. *If $\mathcal{A}$ is saturated, then for all k, every simple expansion $(\mathcal{A}, a_1, \dots, a_n)$ of $\mathcal{A}$ with finitely many elements $a_1, \dots, a_n \in A$ realizes all its k-types.*

Proof. We argue by induction on k. Suppose that $\tau = \tau(x_0, \dots, x_k)$ is a $(k+1)$-type of a simple expansion $\mathcal{A}'$ of $\mathcal{A}$ with finitely many elements. Let $\sigma = \sigma(x_0, \dots, x_k)$ be the set of all finite conjunctions of formulas from τ. Then σ is a $(k+1)$-type as well. Consider $\{\exists x_0 \varphi \mid \varphi \in \sigma\}$. This is a k-type of $\mathcal{A}'$ and by induction hypothesis it is satisfied by elements $a_1, \dots, a_k \in A$. Now $\sigma(x_0, a_1, \dots, a_k)$ is a type of $(\mathcal{A}', a_1, \dots, a_k)$. But, $\mathcal{A}'$ is saturated as well. Thus, $\sigma(x_0, a_1, \dots, a_k)$ is satisfied by an element $a \in A$. But then, $(a, a_1, \dots, a_k)$ satisfies τ. $\dashv$

4.32 Proposition. *Every two equivalent saturated models are partially isomorphic.*

Proof. Suppose that $\mathcal{A}$ and $\mathcal{B}$ are two equivalent saturated models. The strategy of Sy consists in taking care that after the n-th pair of moves (a_n, b_n), he obtains $(\mathcal{A}, a_1, \dots, a_n) \equiv (\mathcal{B}, b_1, \dots, b_n)$. To see that he has a next move, assume that Di plays $a_{n+1} \in A$. Consider the type

$$\tau(x) := \{\varphi \mid (\mathcal{A}, a_1, \dots, a_n) \models \varphi[a_{n+1}]\}$$

of a_{n+1} in $(\mathcal{A}, a_1, \dots, a_n)$. By hypothesis, τ is a type of $(\mathcal{B}, b_1, \dots, b_n)$. As a counter-move of Sy, take b_{n+1} satisfying this type. $\dashv$

4.33 Proposition. *Every countable saturated model has an equivalent that is prime.*

Proof. Omit all non-principal types of elements of the model. $\dashv$

4.34 Uniqueness of countable saturated models. *Every two countable equivalent saturated models are isomorphic.*

Proof. Immediate from Proposition 4.32 and Theorem 3.48. $\dashv$

The following is mainly concerned with types that are maximal. A few basic properties of maximal types are listed in the next lemma.

4.35 Lemma. *Let $\mathcal{A}$ be a model.*

1. *Every type of $\mathcal{A}$ extends to a maximal type.*
2. *The type $\tau = \tau(x_1, \dots, x_n)$ of $\mathcal{A}$ is maximal iff for all $\varphi = \varphi(x_1, \dots, x_n)$, $\varphi \in \tau$ or $\neg\varphi \in \tau$.*
3. *If $\tau(x, a_1, \dots, a_n)$ is a maximal type of $(\mathcal{A}, a_1, \dots, a_n)$, then $\tau(x, x_1, \dots, x_n)$ is a maximal type of $\mathcal{A}$.*

4. *$\mathcal{A}$ is saturated iff every simple expansion of $\mathcal{A}$ with finitely many elements realizes all its maximal 1-types.*

Proof. Exercise 146. $\dashv$

4.36 Existence of Countable Saturated Models. *If a complete theory has only countably many maximal types, then it has a countable saturated model.*

Proof. Suppose that T is a complete theory that only has countably many maximal types. It suffices to show that every countable model of T has a countable saturated elementary extension. Thus, let $\mathcal{A}_0$ be a countable model of T. Construct an elementary chain $\mathcal{A}_0 \prec \mathcal{A}_1 \prec \mathcal{A}_2 \prec \cdots$ of countable models such that for all $n \in \mathbb{N}$:

if $\tau(x, y_1, \ldots, y_k)$ is a maximal type of T and $a_1, \ldots, a_k \in A_n$ are such that $\tau(x, a_1, \ldots, a_k)$ is a type of $(\mathcal{A}_n, a_1, \ldots, a_k)$, then $\tau(x, a_1, \ldots, a_k)$ is realized in $(\mathcal{A}_{n+1}, a_1, \ldots, a_k)$.

Such a construction is possible by assumption, Compactness, and Downward Löwenheim-Skolem Theorem. Now the limit $\mathcal{A} := \bigcup_n \mathcal{A}_n$ of this chain is countable and saturated: if $a_1, \ldots, a_k \in A$ and $\tau(x, a_1, \ldots, a_k)$ is a maximal type of $(\mathcal{A}, a_1, \ldots, a_k)$, then, for some $n \in \mathbb{N}$, $a_1, \ldots, a_k \in A_n$ and $\tau(x, a_1, \ldots, a_k)$ is a maximal type of $(\mathcal{A}_n, a_1, \ldots, a_k)$. Thus, $\tau(x, y_1, \ldots, y_k)$ is a maximal type of T. By construction, $\tau(x, a_1, \ldots, a_k)$ is realized in $(\mathcal{A}_{n+1}, a_1, \ldots, a_k)$ and in $(\mathcal{A}, a_1, \ldots, a_k)$. $\dashv$

A remarkable result using lots of the previous material is the following.

4.37 Vaught's Theorem. *No complete theory has (up to isomorphism) exactly two countable models.*

Proof. Suppose that $\mathcal{A}$ and $\mathcal{B}$ are the only two countable models of the complete theory T. Then since every type of T is realized in a countable model, T cannot have uncountably many maximal (non-principal) types. So, T has a countable model omitting all non-principal types; say, $\mathcal{A}$. Also, by Theorem 4.36, T has a countable saturated model. This cannot be $\mathcal{A}$ as then, by Theorem 4.28, T would be $\aleph_0$-categorical. So, this must be $\mathcal{B}$. Since $\mathcal{A} \not\cong \mathcal{B}$, $\mathcal{B}$ realizes some non-principal type τ. Suppose that $b \in B$ satisfies τ. Consider $T' := Th((\mathcal{B}, b))$. T' has the saturated model $(\mathcal{B}, b)$. Therefore, it cannot have uncountably many maximal types. Therefore, it has a countable model omitting all non-principal types; say, $(\mathcal{C}, c)$. $\mathcal{C}$ cannot be $\cong \mathcal{A}$ since it realizes τ. Thus, $\mathcal{C} \cong \mathcal{B}$ and $(\mathcal{C}, c)$ is saturated. Since T' has a model $(\mathcal{C}, c)$ that is both saturated and omits all non-principal types, all types of T' are principal, and therefore (again by Theorem 4.28) T' has finitely many n-types for each n. However, τ is a non-principal n-type of T, so (by Theorem 4.28 or its proof) T has infinitely many maximal n-types. And every type of T extends to a type of T'. $\dashv$

See Exercise 147 for a complete theory with exactly three countable models.

Exercises

143 Show:

1. Finite models are saturated.
2. The linear orderings η and λ are saturated.
3. The linear orderings ω, ζ and $\omega + \zeta$ are not saturated.
4. The models of an $\aleph_0$-categorical theory are saturated.
5. Linear orderings of type $\zeta \cdot \omega$ realize all their types but are not saturated; orderings of type $\zeta \cdot \eta$ are saturated.

144 (Compare Exercise 140.) Suppose that $\mathcal{A} \equiv \mathcal{B}$, where $\mathcal{A}$ is countable and $\mathcal{B}$ saturated. Prove that $\mathcal{A}$ can be elementarily embedded into $\mathcal{B}$.

145 Show that the following models do not have countable saturated equivalents. In fact, every saturated equivalent of one of these models must have cardinality at least $2^{\aleph_0}$.

1. The simple expansion of the linear ordering η of the rationals with all rationals as constants,
2. the model of arithmetic $(\mathbb{N}, +, \times)$.

Hint. 1. For every irrational r, the type $\tau_r := \{q < x \mid q < r\} \cup \{x < q \mid r < q\}$ is finitely satisfiable in this model. However, if $r_1 < r_2$ are irrationals, then (since for some rational q, $r_1 < q < r_2$) τ_{r_1} and τ_{r_2} cannot be realized by the same element.
2. For every set X of primes, consider the type $\{$"p divides x"$\mid p \in X\} \cup \{$"p does not divide x"$\mid p$ is a prime not in $X\}$.

146 Prove Lemma 4.35.

147 Let $\mathcal{A} = (\mathbb{Q}, <, n)_{n \in \mathbb{N}}$ and $T = Th(\mathcal{A})$.
Show that $\mathcal{B} = (\mathbb{Q}, <, -\frac{1}{n+1})_{n \in \mathbb{N}}$ and $\mathcal{C} = (\mathbb{Q}, <, q_n)_{n \in \mathbb{N}}$ (where $\{q_n\}_{n \in \mathbb{N}}$ is a strictly ascending sequence of rationals that converges to some irrational) are two more countable models of T.
Show that, up to isomorphism, these are the only countable models of T. Which one is saturated? Which one is prime?

148 Show: if a theory has, up to isomorphism, only countably many countable models, then it has a saturated model.

4.6 Recursive Saturation

The Uniqueness Theorem 4.34 is the main reason for the usefulness of countable saturated models. Unfortunately, not every satisfiable theory has a countable saturated model, even if its vocabulary is countable: see Exercise 145 for some examples. Fortunately, there is a useful alternative to saturation: *recursive saturation*. For this notion to make sense, it is easiest to assume from now on that all vocabularies are *finite.* (More generally, you can allow countably infinite vocabularies but require them to be *recursively presented.* This means that you can decide, for each non-logical symbol, whether it is a constant, function, or relation symbol, and in the latter two cases you must be able to compute its arity.) Since we shall apply Ehrenfeucht games, usually you need to assume them not to contain function symbols (though several results below are true for other vocabularies).

The notion of recursive saturation is obtained from ordinary saturation simply by restricting to *computable types.* For the notion of a computable sequence of formulas, see Definition A.13 (page 102).

4.38 Recursive Saturation. A model $\mathcal{A}$ is *recursively saturated* if for every computable sequence of formulas

$$\tau(x_0,\dots,x_n) = \{\varphi_i(x_0,\dots,x_n) \mid i \in \mathbb{N}\}$$

and $a_1,\dots,a_n \in A$: if for every $n \in \mathbb{N}$, $\mathcal{A} \models \exists x_0 \bigwedge_{i<n} \varphi_i[a_1,\dots,a_n]$, then $\tau(x_0,a_1,\dots,a_n)$ is satisfiable in $(\mathcal{A},a_1,\dots,a_n)$.

The above seems to define *recursively enumerable* (or *computable*) saturation instead of *recursive* (or *decidable*) saturation. However, every computable sequence of formulas $\{\varphi_i \mid i \in \mathbb{N}\}$ is logically equivalent with the set $\{\bigwedge_{i<n} \varphi_i \mid n \in \mathbb{N}\}$, which happens to be decidable. This observation is known as *Craig's trick.* (For the proof, see Exercise 150.) Therefore, these notions actually amount to the same thing.

Note that if $\tau = \tau(x_0,a_1,\dots,a_n) = \{\varphi_i(x_0) \mid i \in \mathbb{N}\}$, then the condition of Definition 4.38 becomes (putting $\psi_i := \neg\varphi_i$)

$$\forall a \in A\, \exists i\, \mathcal{A} \models \psi_i[a] \;\Rightarrow\; \exists n\, \forall a \in A\, \exists i < n\, \mathcal{A} \models \psi_i[a].$$

In the following, it usually does not matter a great deal whether you know exactly what computable or recursive enumerability is about. What is used is, that

1. there are but countably many computable types (since the number of computer programs is countable),
2. certain simple types (for instance, singletons) are computable, and
3. certain simple operations applied to computable types produce computable types.

As an example of (3), by the argument of Lemma 4.31, again it does not matter whether in the definition you replace the one variable x_0 by a finite sequence. There are but two places where more is required than just (1)–(3).

4.39 Existence of countable recursively saturated models. *Every satisfiable theory has a countable recursively saturated model.*

First proof. Let T be a satisfiable theory in the countable vocabulary L. Simply extend L by adding a countably infinite set C of new individual constants. Let $\tau_0, \tau_1, \tau_2, \ldots$ enumerate all sets $\tau = \tau(x, \mathbf{c}_1, \ldots, \mathbf{c}_k)$ of $L \cup C$-formulas involving a finite number of constants $\mathbf{c}_1, \ldots, \mathbf{c}_k$ of C where $\tau(x, x_1, \ldots, x_k)$ is a computable type of L. Construct a sequence $T_0 := T, T_1, T_2, \ldots$ by

$$T_{n+1} = \begin{cases} T_n \cup \tau_n(\mathbf{c}) & \text{if this set is satisfiable} \\ T_n & \text{otherwise.} \end{cases}$$

Here, $\tau_n(\mathbf{c})$ is the set obtained from τ by substituting $\mathbf{c}$ for x in every formula. In the first alternative, the witness $\mathbf{c}$ must be chosen not to occur in a formula from $T_n \cup \tau_n$ in order to preserve satisfiability of the resulting union. Now $\bigcup_n T_n$ is maximally satisfiable, has the Henkin property, and the induced canonical model is countable and recursively saturated. Indeed, satisfiability is trivial.

As to maximal satisfiability, suppose that $\bigcup_n T_n \cup \{\varphi\}$ is satisfiable. The singleton $\{\varphi\}$ is a computable type (that has no free variables but may have new constant symbols). Say, $\tau_n = \{\varphi\}$. Then obviously $T_n \cup \tau_n$ is satisfiable, and so $\varphi \in T_{n+1}$.

Henkin: suppose that $\exists x \varphi(x) \in \bigcup_n T_n$. The singleton $\{\varphi(x)\}$ surely is some computable type τ_n. This shows that some instance $\varphi(\mathbf{c})$ is in T_{n+1}.

Finally, the canonical model $\mathcal{A}$ is countable by construction. It is also recursively saturated: if $\tau(x_0, \ldots, x_n)$ is computable and $a_1, \ldots, a_n$ is a sequence of elements in this model, then, by Exercise 100, there are constant symbols $\mathbf{c}_i$ such that $a_i = \mathbf{c}_i^{\mathcal{A}}$. For some n, $\tau_n = \tau(x_0, \mathbf{c}_1, \ldots, \mathbf{c}_n)$. If this is finitely satisfiable in $\mathcal{A}$, then clearly $T_n \cup \tau_n(\mathbf{c})$ ($\mathbf{c}$ chosen as indicated) is satisfiable; hence $\mathbf{c}^{\mathcal{A}}$ satisfies this type in $\mathcal{A}$. ⊣

Second proof. By means of the following lemma.

4.40 Lemma. *Every countable model has a countable recursively saturated elementary extension.*

Proof. A slight modification of the proof for Theorem 4.36 (page 74) suffices. Let $\mathcal{A}$ be a countable model. Using Compactness and Downward Löwenheim-Skolem, construct an elementary chain $\mathcal{A}_0 = \mathcal{A} \prec \mathcal{A}_1 \prec \mathcal{A}_2 \prec \cdots$ of countable models such that for every n, every computable type $\tau(x, y_1, \ldots, y_k)$ and every $a_1, \ldots, a_k \in A_n$, if $\tau(x, a_1, \ldots, a_k)$ is a type of $(\mathcal{A}_n, a_1, \ldots, a_k)$, then it is realized in $(\mathcal{A}_{n+1}, a_1, \ldots, a_k)$. The limit of this chain is the required model. ⊣

Third proof — a sketch, really. First, some

Terminology. Let $\mathcal{M} = (M, \epsilon)$ be a model of (say, Zermelo-Fraenkel) set theory.

$\mathcal{M}$ is (ω-) *non-standard* if its set of natural numbers is not exhausted by its *standard* integers 0, 1, 2,..., but there are *non-standard* integers as well. In other words, the set of natural numbers of $\mathcal{M}$ is not ordered in type ω. Since for each standard n the sentence $\forall x(x < n \to \bigvee_{i<n} x = i)$ must be satisfied, the non-standard integers come after all the standard ones (and of course, after these, the infinite ordinals of $\mathcal{M}$ are still to follow).

In the proof of Lemma 4.43 below you will see that it is very easy to produce such non-standard models using Compactness.

Here is a simple result about such non-standard models.

4.41 Overspill. *If $\mathcal{M}$ is a non-standard model of set theory in which φ is satisfied by every standard integer, then φ is also satisfied by some non-standard integer.*

Proof. If not, φ defines the set of standard integers of $\mathcal{M}$. Thus,

$$\mathcal{M} \models \varphi(0) \wedge \forall n(\varphi(n) \to \varphi(n+1)).$$

By mathematical induction in $\mathcal{M}$, $\mathcal{M} \models \forall n \varphi(n)$; a contradiction, since $\mathcal{M}$ is non-standard. ⊣

You may never have thought about this before, but all models discussed here "live" in the set theoretic universe $(V, \in)$, which is — *except* for the fact that V is not a set — some giant ZF-model. Just as all models live in $(V, \in)$, you can imagine models living in some other (in particular: non-standard) ZF-model.

Let L be a (finite) vocabulary. There is a straightforward computable translation that transforms any L-formula φ into a set-theoretic formula φ^x with one more free variable x such that if the L-model $\mathcal{A}$ lives in the ZF model $\mathcal{M}$ and α is an $\mathcal{A}$-assignment, then $\mathcal{A} \models \varphi[\alpha]$ iff $\mathcal{M} \models \varphi^{\mathcal{A}}[\alpha]$. (See Exercise 154 for more explanations.)

4.42 Proposition. *Every model that lives in a non-standard model of set theory is recursively saturated.*

Assuming this, the existence theorem follows from one more lemma:

4.43 Lemma. *Every satisfiable theory has a model that lives in a countable non-standard model of set theory.*

Proof. Suppose that T is satisfiable. Then T has a model $\mathcal{A}$ that, by necessity, lives in the set theoretic universe $(V, \in)$. Consider the following set of formulas written in the language of set theory with extra constant symbols $\mathcal{A}$ and $\mathbf{c}$ (the construction is not different from that in Exercise 112, page 58):

$$\begin{aligned} \{\varphi^{\mathcal{A}} \mid \varphi \in T\} \quad &\cup \quad \{\text{axioms of set theory}\} \\ &\cup \quad \{\mathbf{c} \text{ is a natural number}\} \\ &\cup \quad \{\mathbf{c} \neq 0, \mathbf{c} \neq 1, \mathbf{c} \neq 2, \ldots\}. \end{aligned}$$

Every subset of this that contains finitely many inequalities $\mathbf{c} \neq n$ only is satisfied in a "model" $(V, \in, \mathcal{A}, n)$ for a suitable natural number n. In particular, this set is finitely satisfiable. Now, apply Compactness and Downward Löwenheim-Skolem to get the required non-standard model. See Exercise 152. ⊣

For an ultraproduct proof of this result, see Proposition 4.62 (page 87). What is problematic here is hidden in Proposition 4.42.

Proof-sketch for 4.42. Suppose that $\mathcal{A}$ lives in the non-standard model $\mathcal{M}$. Let $\tau(x) = \{\varphi_i(x) \mid i \in \mathbb{N}\}$ be a computable type of $\mathcal{A}$. (For ease of exposition, parameters are suppressed.) That is, for every $n \in \mathbb{N}$:

(1) $$\exists a \in A\, \forall i < n\; \mathcal{A} \models \varphi_i(a).$$

Now, *rewrite this as a statement about* $\mathcal{M}$.

Warning: this needs suitable set-theoretic definitions of τ and the relation $\models$ in $\mathcal{M}$. To obtain such a definition for $\models$, see 1.9 page 7. Here, *suitable* means that these definitions express what they should on arguments that are standard. Computable functions do have such definitions. Suitability of a definition of $\models$ is expressed by the Tarski adequacy requirement.

Looking at (1) as a statement about $\mathcal{M}$, you can apply Overspill 4.41: (1) must hold for some non-standard integer n as well. Let $a \in A$ be an element satisfying (1) for this non-standard n. Since every standard integer is less than n, a satisfies τ in $\mathcal{A}$. $\dashv$

This ends the sequence of proofs for 4.39. $\dashv$

The proof of the main Lemma 4.32 for the Uniqueness Theorem employs types that —depending on the models— possibly are not computable. For recursively saturated models, a weaker result holds, which nevertheless is quite useful. Explaining this needs the notion of a *model pair* of two models $\mathcal{A}_1$ and $\mathcal{A}_2$. This is some complex model, appropriately defined, from which the two components $\mathcal{A}_1$ and $\mathcal{A}_2$ may be retrieved again in the sense of Lemma 4.45. Several adequate definitions of this notion are possible, each with its own merits and drawbacks. One implementation of this idea when no function symbols are around is the following.

4.44 Model Pairs. Suppose that L_1 and L_2 are disjunct vocabularies. Choose new unary relation symbols U_1 and U_2. The *model pair* of the L_1-model $\mathcal{A}_1$ and the L_2-model $\mathcal{A}_2$ is a $L_1 \cup L_2 \cup \{U_1, U_2\}$-model $(\mathcal{A}_1, \mathcal{A}_2)$ with universe $A_1 \cup A_2$, such that A_i is the interpretation of U_i and the submodel with universe A_i of the L_i-reduct of $(\mathcal{A}_1, \mathcal{A}_2)$ is $\mathcal{A}_i$ $(i = 1, 2)$.

The requirement that the vocabularies be disjoint can be lifted. In the case that L_1 and L_2 overlap, in particular, if $L_1 = L_2 = L$, a copy L' of L has to be formed and one of the models has to be considered an L'-model.

The sense in which the component models may be retrieved from the pair is the following. (Compare the translation $\varphi \mapsto \varphi^x$ from L-formulas to set-theoretic formulas considered above.)

4.45 Lemma. *There are computable transformations* ^{1} *and* ^{2} *mapping* L_1*- respectively* L_2*-formulas to* $L_1 \cup L_2 \cup \{U_1, U_2\}$*-formulas such that for all* A_i*-assignments* α *and* L_i*-formulas* φ*,* $(\mathcal{A}_1, \mathcal{A}_2) \models \varphi^i[\alpha]$ *iff* $\mathcal{A}_i \models \varphi[\alpha]$*.* *(*$i = 1, 2$*)*

Proof. The formula $\varphi^i = \varphi^{U_i}$ is obtained by *relativising quantifiers* to U_i, see Exercise 153. $\dashv$

Sometimes, you can manage so that $A_1 = A_2$, and the extra U_i and relativization are not needed to form a model pair version that is suitable.

The drawback, that you cannot use function symbols, can always be lifted in practice, since it is possible to replace functions by their graphs.

Whatever the details of the implementation, the following weakened version of Proposition 4.32 holds.

4.46 Proposition. *If $\mathcal{A}_1$ and $\mathcal{A}_2$ are elementarily equivalent models such that $(\mathcal{A}_1, \mathcal{A}_2)$ is recursively saturated, then $\mathcal{A}_1$ and $\mathcal{A}_2$ are partially isomorphic.*

Proof. Let L be the vocabulary involved. Again, the strategy of Sy is to make sure that, after the n-th pair of moves (a_n, b_n) has been played: $(\mathcal{A}_1, a_1, \ldots, a_n) \equiv (\mathcal{A}_2, b_1, \ldots, b_n)$. To see that Sy has an $(n+1)$-st move, assume that Di plays $a_{n+1} \in A_1$. The argument in the proof of 4.32 is not valid here as the type employed may not be computable. However, there is the following trick, which uses the fact that both models form one recursively saturated pair. Let $\tau = \tau(x_1, \ldots, x_{n+1}, y_1, \ldots, y_{n+1})$ be the computable type

$$\{U_2(y_{n+1})\} \cup \{\varphi^1(x_1, \ldots, x_{n+1}) \to \varphi^2(y_1, \ldots, y_{n+1}) \mid \varphi = \varphi(x_1, \ldots, x_{n+1})\}.$$

Put $\tau' = \tau'(y_{n+1}) := \tau(a_1, \ldots, a_{n+1}, b_1, \ldots, b_n, y_{n+1})$. Every finite subset of τ' is satisfied in the model pair $(\mathcal{A}_1, \mathcal{A}_2)$. (For, if $\Phi^1(a_1, \ldots, a_{n+1})$ is the set of true in $\mathcal{A}_1$ sentences $\varphi^1(a_1, \ldots, a_{n+1})$ that are left-hand side of implications in some finite $\sigma \subset \tau'$, then

$$\mathcal{A}_1 \models \exists x_{n+1} \bigwedge \Phi^1(a_1, \ldots, a_n, x_{n+1}),$$

hence $\mathcal{A}_2 \models \exists x_{n+1} \bigwedge \Phi^2(b_1, \ldots, b_n, x_{n+1})$, and any $b_{n+1} \in A_2$ such that $\mathcal{A}_2 \models \bigwedge \Phi^2(b_1, \ldots, b_{n+1})$ satisfies σ.)

By recursive saturation, let b_{n+1} satisfy τ'. Then

$$(\mathcal{A}_1, a_1, \ldots, a_{n+1}) \equiv (\mathcal{A}_2, b_1, \ldots, b_{n+1}). \quad \dashv$$

Similarly:

4.47 Proposition. *If $\mathcal{A}_1$ and $\mathcal{A}_2$ are models such that every positive sentence true in $\mathcal{A}_1$ is satisfied in $\mathcal{A}_2$ and $(\mathcal{A}_1, \mathcal{A}_2)$ is recursively saturated, then $\mathcal{A}_1$ and $\mathcal{A}_2$ are partially homomorphic.*

4.48 Pseudo-uniqueness of Recursively Saturated Models. *If $\mathcal{A}_1$ and $\mathcal{A}_2$ are elementarily equivalent countable models such that $(\mathcal{A}_1, \mathcal{A}_2)$ is recursively saturated, then $\mathcal{A}_1 \cong \mathcal{A}_2$.*

Proof. Immediate from Proposition 4.46 and Theorem 3.48. $\dashv$

Of course, you also have:

4.49 Proposition. *If $\mathcal{A}_1$ and $\mathcal{A}_2$ are countable models such that every positive sentence true in $\mathcal{A}_1$ is satisfied in $\mathcal{A}_2$ and $(\mathcal{A}_1, \mathcal{A}_2)$ is recursively saturated, then $\mathcal{A}_2$ is a homomorphic image of $\mathcal{A}_1$.*

Proof. Immediate from Propositions 4.47 and 3.50. $\dashv$

Exercises

149 Suppose that $\mathcal{A} \equiv \mathcal{B}$, where $\mathcal{A}$ is countable and where the model pair $(\mathcal{A}, \mathcal{B})$ is recursively saturated. Prove that $\mathcal{A}$ can be elementarily embedded into $\mathcal{B}$.

150 (*Craig's trick.*) Suppose that $\{\varphi_i \mid i \in \mathbb{N}\}$ is a computable enumeration of formulas. Show that the set $\{\bigwedge_{i<n} \varphi_i \mid n \in \mathbb{N}\}$ is decidable; that is, describe a decision procedure for it.

Note that, by this trick, every theory with a computably enumerable axiomatisation also has a decidable axiomatisation.
Hint. Let φ be an arbitrary formula. Suppose it is a conjunction of m conjuncts. To check whether φ is in $\{\bigwedge_{i<n} \varphi_i \mid n \in \mathbb{N}\}$, you only need to generate the first m elements of $\{\varphi_i \mid i \in \mathbb{N}\}$.

(It is a fundamental result of *Recursion Theory* — if not its *raison d'être* — that some computably enumerated sets are not decidable. To decide whether a formula is in it, you may not be able to do better than just compute its enumeration; and if your formula, in fact, happens to be not in it, you will never know for sure.)

151 Show that if $\mathcal{M}$ is a non-standard model of set theory in which φ is satisfied by arbitrarily large standard integers, then φ is satisfied by some non-standard integer as well.

152 The proof of Lemma 4.43 has the defect that (as has been explained in Chapter 1 on page 9) there is no way to talk about satisfaction in the proper class "model" $(V, \in, \mathcal{A})$. Give a proof that does not use this.
Hint. The set of sentences employed in the proof must be *consistent.* By the Completeness Theorem A.10 (or A.9, page 101) for first-order logic, it has a model.

153 The *relativization* φ^U is obtained from φ by replacing quantifications $\forall x\psi$ by $\forall x(U(x) \to \psi)$, and $\exists x\psi$ by $\exists x(U(x) \wedge \psi)$. (Compare this with the trick used in Exercise 52.)

Suppose that $\mathcal{A}$ is an L-model, that $U \notin L$ and that $B \subset A$ is the universe of a submodel $\mathcal{B} \subset \mathcal{A}$. Show that for all L-formulas φ and $\mathcal{B}$-assignments β

$$\mathcal{B} \models \varphi[\beta] \Leftrightarrow (\mathcal{A}, B) \models \varphi^U[\beta]$$

where U is interpreted by B in $(\mathcal{A}, B)$.

154 Give more precise definitions of the notion of *living in* and the trans-

lation $\varphi \mapsto \varphi^x$ (for formulas φ not containing x) such that if the L-model $\mathcal{A}$ lives in the set-theoretic model $\mathcal{M}$ and α is an $\mathcal{A}$-assignment, then $\mathcal{A} \models \varphi[\alpha]$ iff $\mathcal{M} \models \varphi^{\mathcal{A}}[\alpha]$.
Hint. Suppose that, for simplicity, $L = \{\mathbf{r}\}$, where $\mathbf{r}$ is a ternary relation symbol. Choose variables y and z not in φ. Let φ' be the set-theoretic formula obtained from φ by (i) replacing quantifiers $\forall u$ and $\exists v$ by $\forall u \in y$ respectively, $\exists v \in y$, and (ii) replacing atoms $\mathbf{r}(u, v, w)$ by the set-theoretic formula $(u, v, w) \in z$. Now φ^x expresses that for some set y and some ternary relation z over y, x is the model (y, z), whereas φ' holds.

155 Assume that $V \subset A$ is not first-order definable (Definition 3.29) on the L-model $\mathcal{A}$ and that $(\mathcal{A}, V)$ is recursively saturated. Show that $a \in V$ and $b \in A - V$ exist such that $(\mathcal{A}, a) \equiv (\mathcal{A}, b)$.
Hint. Show that the recursive type $\{U(x), \neg U(y)\} \cup \{\varphi(x) \leftrightarrow \varphi(y) \mid \varphi$ is an L-formula$\}$ is finitely satisfiable in $(\mathcal{A}, V)$ (U the symbol for V).

Exercise 155 can be used to prove the following result. A first-order definition over a model is a *parametrical* one if it employs an assignment over the model involved. Thus, $\varphi(x, y_1, \ldots, y_n)$ defines the set $\{a \in A \mid \mathcal{A} \models \varphi[a, a_1, \ldots, a_n]\}$ over $\mathcal{A}$, using the *parameters* $a_1, \ldots, a_n$.

4.50 Chang-Makkai Theorem. *If $V \subset A$ is not parametrically first-order definable on the countable model $\mathcal{A}$ and $(\mathcal{A}, V)$ is recursively saturated, then there exist $2^{\aleph_0}$-many sets $V' \subset A$ such that $(\mathcal{A}, V') \cong (\mathcal{A}, V)$.*

Example. If $\mathcal{A}$ is a proper elementary extension of the standard model of arithmetic, then $\mathbb{N}$ is not parametrically first-order definable in $\mathcal{A}$. Therefore, any countable recursively saturated equivalent of $(\mathcal{A}, \mathbb{N})$ has $2^{\aleph_0}$-many initial elementary submodels.

Exercises

156 ♣ Prove Theorem 4.50.

157 Give an example of a model pair of (equivalent) recursively saturated models that itself is not recursively saturated.
Hint. Choose a model with uncountably many types and two recursively saturated equivalents that do not realize the same ones.

4.7 Applications

The following result is Robinson's *Consistency Theorem.*

4.51 Consistency Theorem. *Suppose that T_1 and T_2 are sets of L_1-respectively L_2-sentences and that $L = L_1 \cap L_2$. If there is no L-sentence φ such that both $T_1 \models \varphi$ and $T_2 \models \neg\varphi$, then $T_1 \cup T_2$ has a $(L_1 \cup L_2)$-model.*

Proof. Assume the conditions of the theorem.
Claim. Without loss of generality, it may be assumed that $T_1 \cap T_2$ is a complete L-theory.

Proof. Define $T_1' := T_1 \cup \{\varphi \mid \varphi$ is an L-sentence s.t. $T_2 \models \varphi\}$. By Compactness and assumption, T_1' is satisfiable. Let $\mathcal{A} \models T_1'$. Put

$$T := Th(\mathcal{A} \mid L),$$

the set of L-sentences true in $\mathcal{A}$. Then $T_1 \cup T$ is satisfiable (by $\mathcal{A}$), $T_2 \cup T$ is satisfiable (otherwise, by Compactness, for some $\varphi \in Th(\mathcal{A} \mid L)$ you would have $T_2 \models \neg\varphi$, hence, $\neg\varphi \in T_1'$), $T = (T_1 \cup T) \cap (T_2 \cup T)$ is complete, and there is no L-sentence φ such that both $T_1 \cup T \models \varphi$ and $T_2 \cup T \models \neg\varphi$.

Now, let $(\mathcal{A}_1, \mathcal{A}_2)$ be a countable recursively saturated model pair such that $\mathcal{A}_1 \models T_1$ and $\mathcal{A}_2 \models T_2$. Since $T_1 \cap T_2$ is a complete L-theory, we have that the L-reducts of $\mathcal{A}_1$ and $\mathcal{A}_2$ are equivalent: $\mathcal{A}_1 \mid L \equiv \mathcal{A}_2 \mid L$. By pseudo-uniqueness, $\mathcal{A}_1 \mid L \cong \mathcal{A}_2 \mid L$. But then the isomorphism can be used to copy (say) the $(L_2 - L)$-structure from $\mathcal{A}_2$ to $\mathcal{A}_1$, thereby expanding $\mathcal{A}_1$ to the required model of $T_1 \cup T_2$. $\dashv$

Theorem 4.51 has two well-known corollaries that were originally obtained independently, using no model theory at all. The first one is the *Interpolation Theorem.*

4.52 Interpolation Theorem. *If* $\models \varphi_1 \to \varphi_2$*, then there exists* φ *(an* interpolant*) such that* $\models \varphi_1 \to \varphi$, $\models \varphi \to \varphi_2$*, and every non-logical symbol of* φ *occurs in both* φ_1 *and* φ_2*.*

Proof. Apply Theorem 4.51 to $T_1 := \{\varphi_1\}$, $T_2 := \{\neg\varphi_2\}$ and their respective vocabularies. An interpolant is the same as a sentence φ in the common part of these vocabularies such that $T_1 \models \varphi$ and $T_2 \models \neg\varphi$. $\dashv$

4.53 Corollary. *Disjoint* Σ^1_1*-classes of models can be separated by an elementary class; every* Δ^1_1*-class of models is elementary.* $\dashv$

The $L \cup \{\mathbf{r}\}$-theory T *defines* the relation symbol *implicitly* if every L-model has at most one $L \cup \{\mathbf{r}\}$-expansion that is a model of T, and it *defines* $\mathbf{r}$ *explicitly* if for some L-formula $\psi = \psi(x_1, \ldots, x_k)$ (the *defining* formula),

$$T \models \forall x_1 \cdots \forall x_k (\mathbf{r}(x_1, \ldots, x_k) \leftrightarrow \psi)$$

(that is: ψ defines the interpretation of $\mathbf{r}$ in every model of T). Long ago, Padoa made the trivial but useful observation that a relation symbol that is not defined *implicitly* cannot be defined *explicitly.* Beth showed that this method of proving non-explicit definability is "complete", i.e.: if T does not define $\mathbf{r}$ explicitly, this can always be demonstrated by giving two models witnessing non-implicit definability of $\mathbf{r}$ by T. This result is known as the *Definability Theorem.*

4.54 Definability Theorem. *If* T *defines* $\mathbf{r}$ *implicitly, then* T *defines* $\mathbf{r}$ *explicitly.*

Proof. Assume that the $L \cup \{\mathbf{r}\}$-theory T defines $\mathbf{r}$ implicitly. Let T' be the $L \cup \{\mathbf{r}'\}$-theory obtained from T by replacing the symbol $\mathbf{r}$ by $\mathbf{r}'$. Then the hypothesis can be rendered as $T \cup T' \models \forall x(\mathbf{r}(x) \leftrightarrow \mathbf{r}'(x))$. Choose a fresh individual constant $\mathbf{c}$. Then we also have $T \cup T' \models \mathbf{r}(\mathbf{c}) \to \mathbf{r}'(\mathbf{c})$. By Compactness, choose a conjunction φ of T-sentences such that

$$\{\varphi, \varphi'\} \models \mathbf{r}(\mathbf{c}) \to \mathbf{r}'(\mathbf{c}).$$

(Of course, φ' is the translated φ.)

Rewriting, you obtain $\models \varphi \wedge \mathbf{r}(\mathbf{c}) \to (\varphi' \to \mathbf{r}'(\mathbf{c}))$. By Interpolation, obtain an $L \cup \{\mathbf{c}\}$-sentence $\psi = \psi(\mathbf{c})$ such that

$$\models \varphi \wedge \mathbf{r}(\mathbf{c}) \to \psi(\mathbf{c}) \tag{2}$$

and

$$\models \psi(\mathbf{c}) \to (\varphi' \to \mathbf{r}'(\mathbf{c})). \tag{3}$$

By (2), $T \models \forall x(\mathbf{r}(x) \to \psi(x))$; by (3), $T \models \forall x(\psi(x) \to \mathbf{r}(x))$. Therefore, ψ explicitly defines $\mathbf{r}$ in T. $\dashv$

Note that if $\mathcal{B}$ is a homomorphic image of $\mathcal{A}$, then every positive sentence true in $\mathcal{A}$ is satisfied by $\mathcal{B}$: positive sentences are *preserved* by homomorphisms. Using the modification E^h of the Ehrenfeucht game that is adequate with respect to positive formulas, you obtain a converse: the *Homomorphism Theorem.*

4.55 Homomorphism Theorem. *Every sentence preserved by homomorphisms has a positive equivalent.*

Proof. Assume that φ is preserved by homomorphisms. Put $P := \{\pi \mid \pi$ is positive and $\models \varphi \to \pi\}$. By Compactness, it suffices to establish $P \models \varphi$. Arguing by contradiction, assume that $P \cup \{\neg\varphi\}$ has a model $\mathcal{B}$. Let $N = N_{\mathcal{B}}$ be the set of sentences $\neg\pi$ where π is a positive sentence *not* satisfied by $\mathcal{B}$.

Claim. $N \cup \{\varphi\}$ is satisfiable.

Proof. Otherwise (by Compactness) $\varphi \models \pi_1 \vee \cdots \vee \pi_n$ for finitely many positive π_i false in $\mathcal{B}$. But then $\pi_1 \vee \cdots \vee \pi_n \in P$, contradicting the assumption on $\mathcal{B}$.

The claim shows that a countable recursively saturated model pair $(\mathcal{A}, \mathcal{B})$ exists such that $\mathcal{B}$ satisfies $P \cup \{\neg\varphi\}$ and $\mathcal{A}$ satisfies $N_{\mathcal{B}} \cup \{\varphi\}$. By Proposition 4.49, $\mathcal{B}$ is a homomorphic image of $\mathcal{A}$. By assumption on φ, it follows that $\mathcal{B} \models \varphi$. Contradiction. $\dashv$

4.56 Resplendency. The L-model $\mathcal{A}$ is called *(strongly) resplendent* if for every computable type $\tau = \tau(x_1, \ldots, x_n)$ in a vocabulary L' that expands L with at most finitely many new symbols, and all $a_1, \ldots, a_n \in A$ the following holds:

> if some elementary extension of $(\mathcal{A}, a_1, \ldots, a_n)$ can be expanded to an $L' \cup \{a_1, \ldots, a_n\}$-model of $\tau(a_1, \ldots, a_n)$, then $(\mathcal{A}, a_1, \ldots, a_n)$ itself can be so expanded.

(In the usual definition of this notion, the type τ is a singleton.)

Note that recursive saturation coincides with the special case of resplendency where $L' - L$ consists of exactly one constant symbol. Thus

4.57 Corollary. *Every resplendent model is recursively saturated.* $\dashv$

The hypothesis of the condition of the definition of resplendency has a number of equivalents, summed up by the following lemma.

4.58 Lemma. *Let $\mathcal{A}$, L, and τ be as in Definition 4.56. Then the following conditions are equivalent:*

1. *some elementary extension of $(\mathcal{A}, a_1, \dots, a_n)$ can be expanded to a model of $\tau(a_1, \dots, a_n)$,*
2. *$ELDIAG(\mathcal{A}) \cup \tau(a_1, \dots, a_n)$ is satisfiable,*
3. *$Th((\mathcal{A}, a_1, \dots, a_n)) \cup \tau(a_1, \dots, a_n)$ is satisfiable,*
4. *for every L-formula $\pi = \pi(x_1, \dots, x_n)$ s.t. $\tau \models \pi$: $\mathcal{A} \models \pi[a_1, \dots, a_n]$.*

Proof. Left as Exercise 161. $\dashv$

4.59 Theorem. *Every countable recursively saturated model is resplendent.*

Proof. In fact, you can make sure that the required expansion is recursively saturated again. (Without this requirement the proof below can be simplified in that you only need to enumerate sentences instead of types.) This is another modification of the Henkin argument.

Using formulation 4.58.4, assume that $\mathcal{A}$ is a countable recursively saturated L-model, that $L \subset L'$, and that T is a computable set of L'-sentences the L-consequences of which are valid in $\mathcal{A}$. (We don't mind the parameters from A in T: if $\mathcal{A}$ is recursively saturated, then so are its simple expansions using finitely many elements.)

You find a recursively saturated L'-expansion of $\mathcal{A}$ satisfying T as follows. Fix an enumeration of all sets $\tau = \tau(x, a_1, \dots, a_k)$ where $\tau(x, y_1, \dots, y_k)$ is a computable set of L'-formulas and $a_1, \dots, a_k \in A$. Construct

$$T_0 = T \subset T_1 \subset T_2 \subset \cdots$$

and

$$A_0 = \emptyset \subset A_1 \subset A_2 \subset \cdots \subset A$$

such that for all n:

1. A_n is finite and T_n is a computable set of $L' \cup A_n$-sentences all $L \cup A_n$-consequences of which are satisfied by the $L \cup A_n$-expansion $(\mathcal{A}, a)_{a \in A_n}$ of $\mathcal{A}$,
2. if $\tau = \tau(x, a_1, \dots, a_k)$ is the n-th set of the enumeration, then either for some $a \in A$, $\tau(a, a_1, \dots, a_k) \subset T_{n+1}$, or for some finite $\tau' \subset \tau$, $\neg \exists x \bigwedge \tau'(x, a_1, \dots, a_k) \in T_{n+1}$.

Assume that this construction can be carried out. Let $T^\star$ be the set of logical $L \cup A$-consequences of some T_n. Then:

a. $T^\star$ is satisfiable (by 1).
b. $T^\star$ is maximally satisfiable.
For, assume that $\varphi \notin T^\star$ is an $L' \cup A$-sentence. Suppose that $\tau = \{x = x \wedge \varphi\}$ is the n-th set. By 2, $\neg\exists x\varphi \in T_{n+1}$; therefore, $\neg\varphi \in T^\star$.
c. $T^\star$ has the Henkin property.
For, if $\exists x\varphi(x) \in T^\star$, consider $\tau = \{\varphi(x)\}$ and use 2.
d. $ELDIAG(\mathcal{A}) \subset T^\star$. (By 1 and b.)

Thus, the canonical model of $T^\star$ is (up to isomorphism) an L'-expansion of $\mathcal{A}$. It is recursively saturated by 2.

To see that you can carry out the construction, let $\tau(x, a_1, \ldots, a_k)$ be the n-th set. Put $B := A_n \cup \{a_1, \ldots, a_k\}$. Consider

$$\sigma(x) := \{\varphi(x) \mid \varphi \text{ is an } L_B\text{-formula such that } T_n \cup \tau(x) \models \varphi(x)\}.$$

$T_n \cup \tau$ is computable. Therefore, by the Completeness Theorem from Appendix A (see Lemma A.14), σ is computable as well. Since $\mathcal{A}$ is recursively saturated, there are two possibilities.

(i). σ is satisfied by some element $a \in A$ in $(\mathcal{A}, b)_{b\in B}$. Define

$$T_{n+1} := T_n \cup \tau(a, a_1, \ldots, a_k)$$

and $A_{n+1} := B \cup \{a\}$. Now the first alternative of 2 above is satisfied. We check that 1 holds as well. Assume that $T_{n+1} \models \psi(a)$ where $\psi(x)$ is an L_B-formula. If $a \notin B$, then $T_n \cup \tau(x) \models \psi(x)$, $\psi(x) \in \sigma$, and $(\mathcal{A}, b)_{b\in A_{n+1}} \models \psi(a)$. If $a \in B$, then we have, nevertheless, that

$$T_n \cup \tau(x) \models (x = a \rightarrow \psi(x)),$$

thus $(x = a \rightarrow \psi(x)) \in \sigma$, and in $(\mathcal{A}, b)_{b\in A_{n+1}}$ the formulas $(a = a \rightarrow \psi(a))$ and $\psi(a)$ are satisfied.

(ii). For some finite $\sigma' \subset \sigma$, $(\mathcal{A}, b)_{b\in B} \models \neg\exists x \bigwedge \sigma'$. By Compactness, choose a finite $\tau' \subset \tau$ such that $T_n \cup \tau'(x) \models \bigwedge \sigma'$. Define

$$T_{n+1} := T_n \cup \{\neg\exists x \bigwedge \tau'\} \text{ and } A_{n+1} := A_n.$$

Now, the second alternative of 2 is satisfied. We check that 1 holds: Assume that $T_{n+1} \models \psi$ where ψ is an L_B-sentence. Then T_n logically implies $\neg\exists x \bigwedge \tau' \rightarrow \psi$ and (by choice of τ') $\neg\exists x \bigwedge \sigma' \rightarrow \psi$. By 1, this is valid in $(\mathcal{A}, a)_{a\in B}$. Thus, $(\mathcal{A}, a)_{a\in B} \models \psi$. $\dashv$

4.60 Lemma. *Every resplendent model lives in some non-standard model of set theory.*

Proof. (Sketch.) Let $\mathcal{A}$ be resplendent. Choose a superset $V \supset A$ of A such that $V - A$ has the same power as A (if A is infinite) or is infinite (otherwise). Then the model pair $(V, \mathcal{A})$ is resplendent as well. By Lemma 4.43, some elementary extension of $\mathcal{A}$ lives in a non-standard model of set theory. By resplendency of $(V, \mathcal{A})$, $\mathcal{A}$ *itself* lives in such a model $(V, \epsilon, \mathcal{A})$. $\dashv$

4.61 Corollary. *For a countable model $\mathcal{A}$, the following are equivalent:*

1. *$\mathcal{A}$ is recursively saturated,*
2. *$\mathcal{A}$ is resplendent,*
3. *$\mathcal{A}$ lives in a non-standard model of set theory.*

Proof. Use Theorem 4.59, Lemma 4.60 and Proposition 4.42. $\dashv$

An argument similar to the one for Lemma 4.60 produces the following result. For the notion of an ω-complete ultrafilter, see Exercise 133 (page 65).

4.62 Proposition. *Every ultraproduct that uses a non-ω-complete ultrafilter lives in a non-standard model of set theory.*

Proof. Let $\mathcal{A} = \prod_F \mathcal{A}_i$, where F is a non-ω-complete ultrafilter. Consider the giant models $V_i = (V, \in, \mathcal{A}_i)$ that simply expand the set-theoretic universe $(V, \in)$ with a constant $\mathcal{A}_i$. These form an ultraproduct $\prod_F (V, \in, \mathcal{A}_i)$ that, by the method of Exercise 133, is seen to be non-standard. Note that $\mathcal{A}$ is (isomorphic to) its element $|\lambda i.\mathcal{A}_i|$. $\dashv$

Here is an application to elementary monotone operators. The relevance for fixed point logic is restricted: the expansion of a resplendent model with just one fixed point need not be resplendent any longer.

4.63 Proposition. *Let $\mathcal{M}$ be a non-standard model of set theory. If Γ is an $\mathcal{M}$-definable monotone operator, then both upward and downward closure ordinals of Γ are at most ω.*

Proof. For instance, consider the least fixed point $\Gamma\!\uparrow$ of Γ. Since $\Gamma\!\uparrow\!\omega \subset \Gamma\!\uparrow$, by Γ-induction it suffices to show that $\Gamma(\Gamma\!\uparrow\!\omega) \subset \Gamma\!\uparrow\!\omega$. Working towards a contradiction, assume that $a \in \Gamma(\Gamma\!\uparrow\!\omega) - \Gamma\!\uparrow\!\omega$. In $\mathcal{M}$, (re)construct the fixed point hierarchy for Γ over all of $\mathcal{M}$'s natural numbers (standard and non-standard). Since Γ is definable over $\mathcal{M}$, it does not matter how you read $\Gamma\!\uparrow\!n$ (as defined over $\mathcal{M}$ or not) for n standard. Let $m \in \mathcal{M}$ be any non-standard integer. Then $m-1$ is non-standard as well. For all standard n: $n < m-1$, and hence $\Gamma\!\uparrow\!n \subset \Gamma\!\uparrow\!(m-1)$; therefore,

$$\Gamma\!\uparrow\!\omega = \bigcup_n \Gamma\!\uparrow\!n \subset \Gamma\!\uparrow\!(m-1)$$

and

$$a \in \Gamma(\Gamma\!\uparrow\!\omega) \subset \Gamma(\Gamma\!\uparrow\!(m-1)) = \Gamma\!\uparrow\!m.$$

Since (by assumption on a) for all standard n: $a \notin \Gamma\!\uparrow\!n$, by Overspill 4.41 there exists a non-standard m such that $a \notin \Gamma\!\uparrow\!m$: a contradiction. $\dashv$

4.64 Corollary. *The Scott rank of a model that lives in a non-standard model of set theory is at most ω.*

Compare the next result to Proposition 4.46.

4.65 Corollary. *If $\mathcal{A} \equiv \mathcal{B}$ and the model pair $(\mathcal{A}, \mathcal{B})$ lives in a non-standard model of set theory, then $\mathcal{A}$ and $\mathcal{B}$ are partially isomorphic.*

Proof. By Proposition 3.52, the set of positions in which Sy has a winning strategy for the infinite game is a greatest fixed point of an operator that is easily seen to be definable. $\dashv$

Example. It is not true that models of Scott rank at most ω always live in a non-standard model of set theory. For instance, the linear ordering ω does not live in a non-standard model but does have Scott rank ω.

Proposition 4.63 has an obvious generalization to the case of set-theoretic models that have non-well-founded ordinals but where non-well-foundedness only starts at an ordinal $> \omega$.

Lindström's Theorem is the following result.

4.66 Lindström's Theorem. *First-order logic is a maximally expressive logic for which Downward Löwenheim-Skolem- and Compactness Theorems hold.*

Proof. Of course, the wording of this result is far from precise. For instance, we have not defined the general notion of a *logic*. Nevertheless, we hope to transmit the gist of the result by the following proof.

Logic:
a scheme $\mathcal{Z}$ that for any vocabulary L determines a set $\mathcal{Z}(L)$ of *sentences* and a *satisfaction relation* $\models$ between L-models and $\mathcal{Z}(L)$-sentences such that

- isomorphic models have the same $\mathcal{Z}$-theory
  (note that for the logics discussed here it is straightforward to generalize Theorem 2.3),
- syntactical transformations familiar from first-order logic, such as renaming of symbols, relativization etc., are defined for $\mathcal{Z}$-sentences as well (and have the usual semantic properties).
  (In fact, what you need is Exercise 154 for $\mathcal{Z}$.)

Furthermore, assume *maximality*:

- $\mathcal{Z}$ accommodates the usual propositional connectives (with their usual semantic properties),
- every first-order sentence is a $\mathcal{Z}$-sentence as well (and has its usual meaning).

By the *Downward Löwenheim-Skolem theorem for $\mathcal{Z}$* is meant that every countable satisfiable set of $\mathcal{Z}$-sentences has a countable model. *Compactness* is taken in the usual formulation.

Examples. $\Sigma_1^1(mon)$ satisfies downward Löwenheim-Skolem and Compactness, but has no negation; infinitary logic with *countable* conjunctions and disjunctions satisfies downward Löwenheim-Skolem but is incompact, and the same goes for fixed-point logic.

4.67 Proposition. *Fixed-point logic satisfies the downward Löwenheim-Skolem Theorem.*

Proof. Suppose that $\mathcal{A}$ satisfies the fixed-point sentence φ. Of course, $\mathcal{A}$ lives in the set theoretic universe $(V, \in)$ which contains all ordinals needed to calculate the fixed point hierarchies up to closure for the operators referred to by φ. Let $V' \prec V$ be a countable elementary subsystem containing $\mathcal{A}$ as an element. The part of $\mathcal{A}$ contained in V' is the required submodel $\mathcal{A}'$ of $\mathcal{A}$ satisfying φ. (Note that the ordinals of V' suffice to build the hierarchies over $\mathcal{A}'$ up to closure, as $V' \prec V$.) $\dashv$

What is shown is the following, assuming $\mathcal{Z}$ to be a logic with the required properties.

Claim. *Every two elementarily equivalent models satisfy the same $\mathcal{Z}$-sentences.*

Corollary. *Every $\mathcal{Z}$-sentence has a first-order equivalent.*

Proof of Corollary. (Compare Exercise 104.) Suppose that Φ is a $\mathcal{Z}$-sentence. Let Γ be the set of its first-order consequences. If you can show that $\Gamma \models \Phi$, then (by $\mathcal{Z}$-compactness) Φ follows from a finite subset $\Delta \subset \Gamma$, and hence Φ has the equivalent $\bigwedge \Delta$. So, assume that $\mathcal{A} \models \Gamma$. Suppose that $\mathcal{A} \models \neg\Phi$. Then $Th(\mathcal{A}) \models \neg\Phi$. For, if $\mathcal{B} \models Th(\mathcal{A})$, then $\mathcal{B} \equiv \mathcal{A}$, and hence $\mathcal{B} \models \neg\Phi$ by the Claim. By $\mathcal{Z}$-compactness, for some finite $\Delta \subset Th(\mathcal{A})$ you have $\Delta \models \neg\Phi$. Thus, $\Phi \models \neg \bigwedge \Delta$, $\neg \bigwedge \Delta \in \Gamma$, $\mathcal{A} \models \neg \bigwedge \Delta$, contradicting $\Delta \subset Th(\mathcal{A})$. $\dashv$

Proof of the Claim. Assume that $\mathcal{A}$ and $\mathcal{B}$ are elementary equivalent L-models, but that Φ is a $\mathcal{Z}$-sentence true in $\mathcal{A}$ but false in $\mathcal{B}$. Since $\mathcal{Z}$ accommodates the required syntactic transformation $\varphi \mapsto \varphi^x$, all of this can be expressed as statements about the "model" $(V, \in, \mathcal{A}, \mathcal{B})$: here, we have truth of all ZF-axioms, all equivalences $\varphi^{\mathcal{A}} \leftrightarrow \varphi^{\mathcal{B}}$ (φ a first-order L-sentence), $\Phi^{\mathcal{A}}$ and $\neg\Phi^{\mathcal{B}}$.

From Compactness and Downward Löwenheim-Skolem applied to this set of $\mathcal{Z}$-sentences, plus a type that enforces a non-standard integer, you obtain a countable non-standard model of set theory in which live elementary equivalent models $\mathcal{A}$ and $\mathcal{B}$ that are distinguished by Φ. Now either via Proposition 4.42 or Corollary 4.65 applied to the model pair $(\mathcal{A}, \mathcal{B})$, by pseudo-uniqueness you obtain that $\mathcal{A} \cong \mathcal{B}$ — contradicting the first basic assumption on $\mathcal{Z}$.

For a more direct proof, see Exercise 165. $\dashv$

Exercises

158 Lyndon's strengthening of the Interpolation Theorem 4.52 states that if $\models \varphi_1 \to \varphi_2$, then there exists an interpolant φ with the extra property that every non-logical relation symbol that occurs positively (negatively) in φ occurs positively (negatively) in both φ_1 and φ_2. Prove this.

The extra requirements on occurrences of relation symbols cannot be extended to constant or function symbols. Construct examples that illustrate this.

Hint. Lyndon's result can be obtained as a corollary to 4.52. Use the fact that if $\mathbf{r}$ only occurs positive in ψ and ψ' is obtained from ψ by replacing $\mathbf{r}$ by the new symbol $\mathbf{r}'$, then (by monotonicity) $\models \psi \wedge \forall x \mathbf{r}'(x) \to \psi'$. Similarly, if $\mathbf{r}$ only occurs negatively, then $\models \psi' \wedge \forall x \neg \mathbf{r}(x) \to \psi$.

159 ♣ Show that every model of the positive logical consequences of a theory is elementary submodel of the homomorphic image of a model of that theory.

160 In the context of fixed-point logic it was remarked that if φ is an $\mathbf{r}$-positive $L\cup\{\mathbf{r}\}$-sentence, $\mathcal{A}$ an L-model and $S_1 \subset S_2 \subset A$, then $(\mathcal{A}, S_1) \models \varphi$ implies $(\mathcal{A}, S_2) \models \varphi$.

Now assume that φ is an $L \cup \{\mathbf{r}\}$-sentence with this preservation property. Show that it has an $\mathbf{r}$-positive equivalent.

Hint. For a start, use the game $E^{\mathbf{r}-pos}$ and Proposition 3.51.

161 Prove Lemma 4.58.

162 Let T be an $L \cup \{\mathbf{r}\}$-theory with the property that for every model $(\mathcal{A}, R) \models T$ (where $\mathcal{A}$ is an L-model and R interprets $\mathbf{r}$), every automorphism of $\mathcal{A}$ is an automorphism of $(\mathcal{A}, R)$ as well. Show that a finite set Φ of L-formulas exists such that whenever $(\mathcal{A}, R) \models T$, then R is defined on $\mathcal{A}$ by some formula from Φ.

Hint. Apply Exercise 155.

163 Assume that $B \subset A$ is first-order definable in the resplendent model $\mathcal{A}$. Show that B is finite or $|B| = |A|$.

164 Show that every infinite resplendent linearly ordered model embeds η.

165 Complete the following direct proof of Lindström's Theorem, that is: assuming that $\mathcal{A} \equiv \mathcal{B}$, show that $\mathcal{A}$ and $\mathcal{B}$ satisfy the same $\mathcal{Z}$-sentences (under suitable conditions on the logic $\mathcal{Z}$).

Suppose that $\mathcal{A} \equiv \mathcal{B}$ and, moreover, that the $\mathcal{Z}$-sentence Φ is satisfied in $\mathcal{A}$ but not in $\mathcal{B}$. For n a natural number, consider the expanded model pair $\mathcal{M}_n = (\mathcal{A}, \mathcal{B}, R_0, \ldots, R_n)$, where the $2i$-ary relations $R_i \subset A^i \times B^i$ $(i = 0, \ldots, n)$ are defined by

$$R_i(a_1, \ldots, a_i, b_1, \ldots, b_i) :\equiv (\mathcal{A}, a_1, \ldots, a_i) \equiv^{n-i} (\mathcal{B}, b_1, \ldots, b_i).$$

The model $\mathcal{M}_n$ satisfies sentences expressing that

1. Φ is true in $\mathcal{A}$ but false in $\mathcal{B}$,
2. if $i \leq n$ and $R_i(a_1, \ldots, a_i, b_1, \ldots, b_i)$ holds, then $\{(a_1, b_1), \ldots, (a_i, b_i)\}$ is a local isomorphism between $\mathcal{A}$ and $\mathcal{B}$,
3. a. R_0 (which has 0 arguments) is true (of the empty sequence),
   b. if $i < n$ and $R_i(a_1, \ldots, a_i, b_1, \ldots, b_i)$ holds, then for all $a \in A$ there exists $b \in B$ such that $R_{i+1}(a_1, \ldots, a_i, a, b_1, \ldots, b_i, b)$, and vice versa.

By the Downward Löwenheim-Skolem and Compactness Theorem for $\mathcal{Z}$, there is a *countable* complex $(\mathcal{A}, \mathcal{B}, R_0, R_1, R_2, \ldots)$ with an *infinite* sequence $R_0, R_1, R_2, \ldots$ that satisfies these requirements for *every* i. By requirements 2 and 3 it follows that $\mathcal{A} \cong \mathcal{B}$. However, this contradicts requirement 1.

Bibliographic remarks

The proof of Compactness is a modification of the Henkin proof of the Completeness Theorem from Appendix A.

Exercise 114 is due to Szpilrajn, Exercise 113 to Erdös and de Bruijn. Results such as these can already be carried out using Compactness for the propositional calculus. The crucial set of propositional formulas for Exercise 113 in the context of *finite* graphs is instrumental in obtaining a polynomial reduction of the graph-colorability problem to propositional satisfiability.

The notion of a Horn formula is due to Alfred Horn (On sentences which are true of direct unions of algebras, *Journal of Symbolic Logic* 16:14–21), though his notion is more general. The definition used here comes from logic programming theory.

Theorem 4.17 goes back to Łos' (1955).

Two sources for Theorem 4.28 are Ryll-Nardzewski 1959 and Svenonius 1959.

A elegant introduction to 0–1 laws is Gurevich 1992. An original source is Fagin 1976. For the 0–1-law for fixed-point logic, see, for instance, Blass et al. 1985.

On the class of finite models with ordering, the subclasses that have a polynomial time decision problem coincide with those definable by a fixed point formula. (Fagin and Immerman, independently.)

For the curious history of recursively saturated models (via infinitary admissible languages), see Barwise and Schlipf 1976. This is also the source for the notion of resplendency.

The Consistency Theorem 4.51 is due to A. Robinson.

The Interpolation Theorem 4.52 is from Craig 1957. For the Definability Theorem 4.54, see Beth 1953. The three results were found independently and proved by quite different means. (Craig used proof theory and Beth employed his tableaux.)

The Homomorphism Theorem 4.55 is due to Lyndon 1959b.

Theorem 4.59 is due to Ressayre 1977.

Theorem 4.66 is from Lindström 1969. The proof in the text is in the spirit of Friedman's rediscovery of this result; the one indicated in Exercise 165 is closer to the original. A general reference for the abstract notion of a logic is Barwise 1985.

The Lyndon Interpolation Theorem from Exercise 158 is from Lyndon 1959a.

The result in Exercise 162 is due to Svenonius.

A

Deduction and Completeness

The subject of this appendix, the Completeness Theorem for first-order logic (with respect to a system of natural deduction), does not properly belong to model theory. The reasons to include it here are (i) the method of proof is a simple modification of the one used in Section 4.1, and (ii) its addition makes the book more self-contained as a first-order logic text.

The Completeness Theorem and its companion, the Soundness Theorem, show that there is an adequate combinatorial approach to the notion of logical consequence that is close to the usual notion of a mathematical proof. Cf. A.12 for a further elaboration of the meaning of these results.

Completeness and soundness show that a (first-order) sentence follows logically from certain (first-order) sentences if and only if it has a *proof* from those sentences, in a suitable system of deduction. This appendix explains a system of *natural deduction*, which is based on the following choice of logical primitives.

- $\bot$ (*falsum*), which stands for a (by definition) false sentence,
- $\rightarrow$ and $\wedge$, implication and conjunction,
- $\forall$, the universal quantifier.

Other logical constants are assumed to be defined from these, by

$$\begin{aligned} \neg\varphi &:= \varphi \rightarrow \bot \\ \varphi \leftrightarrow \psi &:= (\varphi \rightarrow \psi) \wedge (\psi \rightarrow \varphi) \\ \varphi \vee \psi &:= \neg\varphi \rightarrow \psi \\ \exists x\varphi &:= \neg\forall x\neg\varphi. \end{aligned}$$

Formally, a *derivation* is a tree of formulas, constructed according to certain *derivation rules*. Every logical constant has its own rules; usually, an *introduction* and an *elimination* rule. The root of such a tree is the *conclusion* of the derivation, that is: the formula proved by it. The leaves of the tree are its *hypotheses* — except when such a leaf has been *discharged*.

A.1 Rules of Natural Deduction

The following table visually presents the rules for $\rightarrow, \wedge, \bot$ and $\forall$. In them, the derived formula has been put under a horizontal bar. Discharge of hypotheses (in the $\bot$-rule and the $\rightarrow$-introduction rule) is indicated by putting the formula between square brackets [and].

	introduction	elimination
$\bot$	none	$\begin{array}{c}[\varphi \rightarrow \bot]\\ \vdots \\ \bot \\ \hline \varphi\end{array}$
$\rightarrow$	$\begin{array}{c}[\varphi]\\ \vdots \\ \psi \\ \hline \varphi \rightarrow \psi\end{array}$	$\dfrac{\varphi \quad \varphi \rightarrow \psi}{\psi}$
$\wedge$	$\dfrac{\varphi \quad \psi}{\varphi \wedge \psi}$	$\dfrac{\varphi \wedge \psi}{\varphi} \quad \dfrac{\varphi \wedge \psi}{\psi}$
$\forall$	$\dfrac{\varphi(x)}{\forall x\, \varphi(x)}$	$\dfrac{\forall x\, \varphi(x)}{\varphi(t)}$

The precise definition of what constitutes a derivation involves the notion of an *identity axiom.*

A.1 Identity Axioms. These are all formulas of one of the following shapes. Here, t, s, u, $s_1, \ldots, s_n$, t_1, $\ldots$, t_n are arbitrary terms (variables allowed), $\mathbf{f}$ is an n-ary function symbol and $\mathbf{r}$ is an n-ary relation symbol.

1. $t = t$,
2. $s = t \rightarrow t = s$,
3. $s = t \rightarrow (t = u \rightarrow s = u)$,
4. $s_1 = t_1 \rightarrow (\cdots (s_n = t_n \rightarrow \mathbf{f}(s_1, \ldots, s_n) = \mathbf{f}(t_1, \ldots, t_n)) \cdots)$,
5. $s_1 = t_1 \rightarrow (\cdots (s_n = t_n \rightarrow (\mathbf{r}(s_1, \ldots, s_n) \rightarrow \mathbf{r}(t_1, \ldots, t_n))) \cdots)$.

A.2 Lemma. *Identity axioms are logically valid.*

Proof. This is Exercise 166. $\dashv$

Now follows the precise, *inductive* definition of what constitutes a derivation.

A.3 Derivation, Conclusion, Hypotheses, their Discharge.

(0) A tree consisting of a single formula is a (*rudimentary*) derivation. The conclusion of such a derivation is the formula itself. Similarly, its only hypothesis is the formula itself, *unless this formula is an identity axiom*, in which case the derivation has *no* hypotheses.

($\wedge$E) ($\wedge$-*elimination*) If $\mathcal{D}$ is a derivation with conclusion $\varphi \wedge \psi$, then the two trees obtained from $\mathcal{D}$ by putting φ respectively ψ underneath as a new root are also derivations. The first has conclusion φ, the second has conclusion ψ. The hypotheses of these new derivations are the same as those of $\mathcal{D}$.

($\wedge$I) ($\wedge$-*introduction*) If $\mathcal{D}_1$ and $\mathcal{D}_2$ are derivations with conclusions φ and ψ (or vice versa), then the tree obtained by joining $\mathcal{D}_1$ and $\mathcal{D}_2$ with $\varphi \wedge \psi$ as a new root is also a derivation. The conclusion of this derivation is $\varphi \wedge \psi$. Its hypotheses are those of $\mathcal{D}_1$ plus those of $\mathcal{D}_2$.

($\rightarrow$E) (MP, *Modus Ponens*, $\rightarrow$-*elimination*) If $\mathcal{D}_1$ and $\mathcal{D}_2$ are derivations with conclusions φ respectively $\varphi \rightarrow \psi$ (or vice versa), then the tree obtained by joining $\mathcal{D}_1$ and $\mathcal{D}_2$ with ψ as a new root is also a derivation. Its conclusion is ψ. Its hypotheses are those of $\mathcal{D}_1$ plus those of $\mathcal{D}_2$.

($\rightarrow$I) (D, *Deduction-rule*, $\rightarrow$-*introduction*) If $\mathcal{D}$ is a derivation with conclusion ψ, and φ is an arbitrary formula, then the tree obtained from $\mathcal{D}$ by putting $\varphi \rightarrow \psi$ below it as a new root is also a derivation. The conclusion of this tree is $\varphi \rightarrow \psi$. Its hypotheses are those of $\mathcal{D}$ *minus* φ; φ is said to be *discharged*.

($\perp$) (RAA, *Reductio ad Absurdum*; $\perp$-rule) Suppose that $\mathcal{D}$ is a derivation with conclusion $\perp$ and let φ be an arbitrary formula. The tree obtained from $\mathcal{D}$ by putting φ beneath it as a new root is also a derivation. Its conclusion is φ. Its hypotheses are those of $\mathcal{D}$, *minus* $\neg\varphi$ ($= \varphi \rightarrow \perp$). Again, $\neg\varphi$ is said to be *discharged*.

($\forall$I) (G, *Generalisation*, $\forall$-*introduction*) Let $\mathcal{D}$ be a derivation with conclusion φ and x a variable *that is not free in a hypothesis of* $\mathcal{D}$. The tree obtained from $\mathcal{D}$ by putting $\forall x\varphi$ underneath as a new root is also a derivation. Its conclusion is $\forall x\varphi$. Its hypotheses are the same as those of $\mathcal{D}$.

($\forall$E) (I, *Instantiation*, $\forall$-*elimination*) Let $\mathcal{D}$ be a derivation with conclusion $\forall x\varphi(x)$ and let t be a term *that is substitutable for* x *in* φ. (Cf. Exercise 4, page 7.) The tree obtained from $\mathcal{D}$ by putting $\varphi(t)$ underneath as a new root is also a derivation. Its conclusion is $\varphi(t)$. Its hypotheses are the same as those of $\mathcal{D}$.

Warning for the restrictions on the application of $\forall$-rules. The one on generalisation looks simple, but it is easily overlooked. The reason is that the rule is applied at the bottom of a derivation tree, whereas you have to check whether the relevant variable is not free high up the tree.

A.4 Example Derivations.

1. $\varphi = \varphi(z)$, y is not free in $\varphi(z)$ and is substitutable for z in φ. A derivation of $\forall y\varphi(y)$ with hypothesis $\forall z\varphi(z)$:

$$\text{G}\,\dfrac{\text{I}\,\dfrac{\forall z\varphi(z)}{\varphi(y)}}{\forall y\varphi(y)}$$

2. A derivation without hypotheses of $\exists x\forall y(\mathbf{r}(x)\to\mathbf{r}(y))$.
N.B.: $\exists$ has been defined as $\neg\forall\neg$, and $\neg\cdots$ in turn has been defined as $\cdots\to\bot$. The formula derived therefore really is

$$\forall x(\forall y(\mathbf{r}(x)\to\mathbf{r}(y)\to\bot)\to\bot.$$

Hypotheses discharged and the rule that accomplishes the discharging have been labeled by the same number.

$$\text{MP}\,\dfrac{\text{G}\,\dfrac{\text{D–1}\,\dfrac{\text{RAA}\,\dfrac{\text{MP}\,\dfrac{[\neg\mathbf{r}(x)]^2 \qquad [\mathbf{r}(x)]^1}{\bot}}{\mathbf{r}(y)}}{\mathbf{r}(x)\to\mathbf{r}(y)}}{\forall y(\mathbf{r}(x)\to\mathbf{r}(y))} \qquad \dfrac{[\forall x(\forall y(\mathbf{r}(x)\to\mathbf{r}(y))\to\bot)]^3}{\forall y(\mathbf{r}(x)\to\mathbf{r}(y))\to\bot}\,\text{I}}{\bot}$$

$$\text{D–3}\,\dfrac{\text{MP}\,\dfrac{\text{G}\,\dfrac{\text{D}\,\dfrac{\text{I}\,\dfrac{\text{G}\,\dfrac{\text{RAA–2}\,\dfrac{\bot}{\mathbf{r}(x)}}{\forall x\mathbf{r}(x)}}{\mathbf{r}(y)}}{\mathbf{r}(x)\to\mathbf{r}(y)}}{\forall y(\mathbf{r}(x)\to\mathbf{r}(y))} \qquad \dfrac{[\forall x(\forall y(\mathbf{r}(x)\to\mathbf{r}(y))\to\bot)]^3}{\forall y(\mathbf{r}(x)\to\mathbf{r}(y))\to\bot}\,\text{I}}{\bot}}{\forall x(\forall y(\mathbf{r}(x)\to\mathbf{r}(y))\to\bot)\to\bot}$$

This example is rather artificial — but of course, so is the derived formula.

Suppose, furthermore, that the variable x is not free in the formula ψ. Below follow (3a/b) the provable equivalence of

$$\exists x\varphi(x)\to\psi \text{ with } \forall x(\varphi(x)\to\psi),$$

and (3b/c) that of

$$\forall x\varphi(x)\to\psi \text{ with } \exists x(\varphi(x)\to\psi).$$

There is nothing against trying yourself first!

3a. Here follows a derivation of $\exists x\varphi(x)\to\psi$ that uses the formula

$$\forall x(\varphi(x)\to\psi)$$

as a hypothesis.

$$\text{D–3}\,\dfrac{\text{RAA–2}\,\dfrac{\text{MP}\,\dfrac{\text{G}\,\dfrac{\text{D–1}\,\dfrac{\text{MP}\,\dfrac{\text{MP}\,\dfrac{\text{I}\,\dfrac{\forall x(\varphi(x)\to\psi)}{\varphi(x)\to\psi} \quad [\varphi(x)]^1}{\psi} \quad [\neg\psi]^2}{\bot}}{\neg\varphi(x)}}{\forall x\neg\varphi(x)} \quad [\neg\forall x\neg\varphi(x)]^3}{\bot}}{\psi}}{\neg\forall x\neg\varphi(x)\to\psi}$$

3b. Likewise, a derivation of $\forall x(\varphi(x)\to\psi)$ with hypothesis $\exists x\varphi(x)\to\psi$:

$$\text{G}\,\dfrac{\text{D–2}\,\dfrac{\text{MP}\,\dfrac{\text{D–1}\,\dfrac{\text{MP}\,\dfrac{\text{I}\,\dfrac{[\forall x\neg\varphi(x)]^1}{\neg\varphi(x)} \quad [\varphi(x)]^2}{\bot}}{\neg\forall x\neg\varphi(x)} \quad \neg\forall x\neg\varphi(x)\to\psi}{\psi}}{\varphi(x)\to\psi}}{\forall x(\varphi(x)\to\psi)}$$

3c. Likewise, a derivation of $\forall x\varphi(x)\to\psi$ with hypothesis $\exists x(\varphi(x)\to\psi)$.

$$\text{G}\,\dfrac{\text{D–2}\,\dfrac{\text{D–1}\,\dfrac{\text{MP}\,\dfrac{\text{I}\,\dfrac{[\forall x\varphi(x)]^1}{\varphi(x)} \quad [\varphi(x)\to\psi]^2}{\psi}}{\forall x\varphi\to\psi}}{(\varphi\to\psi)\to(\forall x\varphi\to\psi)}}{\forall x((\varphi\to\psi)\to(\forall x\varphi\to\psi))}$$

This derivation is completed by derivation 3a (where the roles of φ and ψ from 3a now are being played by $\varphi\to\psi$ and $\forall x\varphi\to\psi$), plus one application of modus ponens, involving the hypothesis.

3d. Likewise, a derivation of $\exists x(\varphi(x)\to\psi)$ with hypothesis $\forall x\varphi(x)\to\psi$. Start as follows.

$$\text{G}\,\dfrac{[\forall x\neg(\varphi\to\psi)]^0}{\neg(\varphi\to\psi)}$$

Using only propositional rules, first pursue with a derivation of φ, and next with a derivation of $\neg\psi$:

$$\text{RAA–1}\,\dfrac{\text{MP}\,\dfrac{\text{D–2}\,\dfrac{\text{D}\,\dfrac{\text{MP}\,\dfrac{[\neg\varphi]^1 \qquad [\varphi]^2}{\bot}}{\psi}}{\varphi\to\psi} \qquad \begin{matrix}\vdots\\ \neg(\varphi\to\psi)\end{matrix}}{\bot}}{\varphi}
\qquad\qquad
\text{RAA–1}\,\dfrac{\text{MP}\,\dfrac{\text{D}\,\dfrac{[\psi]^1}{\varphi\to\psi} \qquad \begin{matrix}\vdots\\ \neg(\varphi\to\psi)\end{matrix}}{\bot}}{\neg\psi}\;.$$

These newly derived formulas allow the following closing-off (which is not a derivation in itself, so the application of generalisation is allowed!):

$$\text{D–0}\,\dfrac{\text{MP}\,\dfrac{\text{MP}\,\dfrac{\text{G}\,\dfrac{\begin{matrix}\vdots\\ \varphi\end{matrix}}{\forall x\varphi} \qquad \forall x\varphi\to\psi}{\psi} \qquad \begin{matrix}\vdots\\ \neg\psi\end{matrix}}{\bot}}{\neg\forall x\neg(\varphi\to\psi)}$$

Exercises

166 Prove Lemma A.2.

167 Because of the restriction on the Instantiation Rule ∀E, the following configuration is not a derivation

$$\text{I}\,\dfrac{\forall x\forall y\varphi(x,y)}{\forall y\varphi(y,y)}$$

Produce a correct derivation with conclusion $\forall y\varphi(y,y)$ and hypothesis $\forall x\forall y\varphi(x,y)$.
Hint. You may need *five* steps.

168 Produce derivations without hypotheses for the following list of formulas.

1. $\exists x\forall y\Phi \longrightarrow \forall y\exists x\Phi$,
2. $\forall x(\Phi\to\Psi) \longrightarrow (\forall x\Phi\to\forall x\Psi)$; $\forall x(\Phi\to\Psi) \longrightarrow (\exists x\Phi\to\exists x\Psi)$,
3. $\forall x\forall y\Phi \longrightarrow \forall y\forall x\Phi$; $\exists x\exists y\Phi \longrightarrow \exists y\exists x\Phi$,
4. $\forall x(\Phi\wedge\Psi) \longleftrightarrow (\forall x\Phi\wedge\forall x\Psi)$.

169 Produce derivations without hypotheses for the following formulas.

1. $\forall x\exists y(\mathbf{r}(x)\to\mathbf{r}(y))$,
2. $\forall x\exists y(\mathbf{r}(y)\to\mathbf{r}(x))$,
3. $\exists x\forall y(\mathbf{r}(y)\to\mathbf{r}(x))$,
4. $\forall x\neg\forall y(\mathbf{r}(y,x)\leftrightarrow\neg\mathbf{r}(x,x))$.

Hint. As to 3, cf. Example A.4.2.

A.5 Derivable From. The formula φ is *derivable from* the set of formulas Γ if there is a derivation of which φ is the conclusion, whereas all hypotheses are members of Γ. Notation: $\Gamma\vdash\varphi$.

Notations like $\vdash \varphi$, $\varphi_1, \ldots, \varphi_n \vdash \varphi$, and $\Gamma, \psi \vdash \varphi$ are used to indicate that $\emptyset \vdash \varphi$, $\{\varphi_1, \ldots, \varphi_n\} \vdash \varphi$ and $\Gamma \cup \{\psi\} \vdash \varphi$, respectively.

Strictly speaking, this definition is *relative to* a given vocabulary. Thus, more precise is: the notation $\Gamma \vdash_L \varphi$ (where L is some vocabulary) presupposes that $\Gamma \cup \{\varphi\}$ is a set of L-formulas, and indicates that a derivation of φ from Γ exists which consists entirely of L-formulas.

Exercises

170 (Elimination rule for $\exists$.)

1. Suppose that $\Gamma, \neg\varphi(x) \vdash \psi$ and that x is not free in ψ or in a formula of Γ. Show that $\Gamma, \neg\forall x\varphi(x) \vdash \psi$.
2. Suppose that $\Gamma, \varphi(x) \vdash \psi$ and that x is not free in ψ or in a formula of Γ. Show that $\Gamma, \exists x\varphi(x) \vdash \psi$.

171 Suppose that the term t is substitutable for x in $\varphi(x)$. Show that $\varphi(t) \vdash \exists x\varphi(x)$ and that $\neg\varphi(t) \vdash \neg\forall x\varphi(x)$.

172 Suppose that $t = t(x)$ and that t_1 and t_2 are substitutable for x in $\varphi(x)$. Show that

1. $t_1 = t_2 \vdash t(t_1) = t(t_2)$,
2. $t_1 = t_2, \varphi(t_1) \vdash \varphi(t_2)$.

173 In Example A.4.1 you find a derivation of $\forall y\varphi(y)$ from $\forall x\varphi(x)$ that involves a generalisation with respect to y. Suppose that $\Gamma \vdash \forall x\varphi(x)$, and that y is substitutable for x in φ. Show that $\Gamma \vdash \forall y\varphi(y)$ — even if y is free in a formula of Γ.

174 Which one of the two sentences

1. $\forall x(\mathbf{r}_1(x) \vee \mathbf{r}_2(x)) \rightarrow (\forall x\mathbf{r}_1(x) \vee \forall x\mathbf{r}_2(x))$,
2. $\forall x\forall y(\mathbf{r}_1(x) \vee \mathbf{r}_2(y)) \rightarrow (\forall x\mathbf{r}_1(x) \vee \forall y\mathbf{r}_2(y))$

is logically valid? Produce a derivation which has no hypothesis.

175 (Prenex Normal Forms.) A *prenex (normal) form* is a formula which begins with a sequence of quantifiers (the *prefix*) after which a quantifier-free formula (the *matrix*) follows. (Since $\exists$ has been defined here as $\neg\forall\neg$ one should in fact say: a prefix is a sequence of universal quantifiers and negation symbols.)

Prenex Normal Form Theorem. *Every formula has an equivalent in prenex normal form.*

(For *equivalent* you can read both *logical* and *provable* equivalence here.) Show this.
Hint. The proof of *provable* equivalence uses Example A.4.3.

A.2 Soundness

A.6 Lemma. *Suppose that $\mathcal{D}$ is a derivation, $\mathcal{A}$ a model and α an $\mathcal{A}$-assignment. If α satisfies every hypothesis of $\mathcal{D}$ in $\mathcal{A}$, then it also satisfies the conclusion of $\mathcal{D}$ in $\mathcal{A}$.*

Proof. Argue by induction with respect to derivations. Here are a few examples.
1. $\mathcal{D} = \{\varphi\}$ is rudimentary.
(i) φ is hypothesis of $\mathcal{D}$. Trivial.
(ii) φ is an identity axiom. By Lemma A.2, identity axioms are logically valid.
2. The last rule that is applied in $\mathcal{D}$ is RAA.
For instance, $\mathcal{D}$ is formed from a subderivation that derives $\bot$ by putting φ underneath and discharging $\neg\varphi$. By induction hypothesis, you may assume that every assignment that satisfies the hypotheses of this subderivation also satisfies $\bot$. This in fact means that no assignment can satisfy all hypotheses of the subderivation, as, by definition, no assignment satisfies $\bot$. Now suppose that the assignment α satisfies all hypotheses of $\mathcal{D}$ in the model $\mathcal{A}$. It cannot satisfy $\neg\varphi$ (since otherwise, all hypotheses of the subderivations are satisfied). Therefore, it satisfies φ.
3. The last rule that has been applied in $\mathcal{D}$ is $\forall$E.
For instance, the conclusion $\varphi(t)$ is drawn by means of $\forall$E from the next-to-last formula $\forall x\varphi(x)$ of $\mathcal{D}$, and t is substitutable for x in φ. According to the induction hypothesis (applied to the subderivation of $\mathcal{D}$ that derives $\forall x\varphi(x)$), $\forall x\varphi(x)$ is satisfied by α in $\mathcal{A}$. Now use that $\forall x\varphi(x) \models \varphi(t)$. (Cf. Exercise 6, page 7.)
4. The last rule applied in $\mathcal{D}$ is $\forall$I.
For instance, the conclusion $\forall x\varphi(x)$ of $\mathcal{D}$ is drawn using $\forall$I from the next-to-last formula $\varphi(x)$ of $\mathcal{D}$, whereas x is not free in a hypothesis of the subderivation of $\mathcal{D}$ that derives $\varphi(x)$. Suppose that α satisfies every hypothesis of $\mathcal{D}$ in $\mathcal{A}$. We want to show that $\mathcal{A} \models \forall x\varphi[\alpha]$. I.e., for an arbitrary element a of the universe of $\mathcal{A}$ it must be shown that $\mathcal{A} \models \varphi[\alpha^x_a]$, where α^x_a is the modification of α that maps x to a (cf. Definition 1.7, page 5). As x is not free in a hypothesis of $\mathcal{D}$, it follows that these hypotheses are satisfied by α^x_a as well (cf. Exercise 2, page 6). An application of the induction hypothesis to α^x_a and the subderivation shows that $\mathcal{A} \models \varphi[\alpha^x_a]$. ⊣

A.7 Soundness Theorem. $\Gamma \vdash \varphi \Rightarrow \Gamma \models \varphi$.

Proof. Immediate from Lemma A.6. ⊣

176 Exercise. Consider the set $\{0, \frac{1}{2}, 1\}$ as a set of truth values ($0 =$ *False*, $1 =$ *True*, $\frac{1}{2} =$ "In between"). Define the truth value of $\varphi \to \psi$ to be 1 if the truth values of φ is $\leq$ that of ψ and let it be the truth value of

ψ otherwise. $\bot$ always obtains value 0; the value of $\varphi \wedge \psi$ is the *minimum* of those of φ and ψ.

Check that the propositional rules remain sound if RAA is modified in such a way that it no longer is capable of discharging hypotheses. I.e.: if $\mathcal{D}$ is a derivation in which only propositional rules have been applied that derives φ from Γ and in which no hypothesis is discharged under an application of RAA, and γ is a truth assignment (allowing $\frac{1}{2}$), then $min\{\gamma(\psi) \mid \psi \in \Gamma\} \leq \gamma(\varphi)$ (where $\gamma(\varphi)$ is the truth value of φ that is calculated by means of γ).

Show that every derivation of $\neg\neg p \rightarrow p$ (where p is any atom not involving $=$) discharges a hypothesis using RAA.

A.3 Completeness

A.3.1 Consistency

A.8 Consistency. A set of sentences Γ is *consistent* if $\Gamma \nvdash \bot$.

Note that from an inconsistent set you can (by RAA) derive every formula.

Also note: the following bears no relation to Theorem 4.51, although both results carry the same name.

A.9 Consistency Theorem. *Every consistent set of sentences has a model.*

Proof. Cf. Subsection A.3.2. $\dashv$

A.10 Completeness Theorem. *A sentence that logically follows from a set of sentences can also be derived from that set:* $\Gamma \models \varphi \Rightarrow \Gamma \vdash \varphi$.

Proof. See Exercise 179. $\dashv$

A.11 Corollary. $\Gamma \models \varphi \Leftrightarrow \Gamma \vdash \varphi$.

Proof. Theorems A.7 and A.10. $\dashv$

A.12 Positive and Feasible Decision Methods. The notion of a logically valid sentence is *positively decidable.* This means: there is an algorithmic method — one that can, at least in principle, be carried out by a computer — that allows you to ascertain logical validity of a formula *in the case that, indeed, the formula is logically valid.* However, the method is of no use when confronted with formulas that are not logically valid. The method is based on Theorem A.11. Thus, if you want to know whether a given sentence happens to be logically valid, all you need to do is to generate (or to have generate) in some systematic order, all possible derivations and to check whether one of them constitutes a hypothesis-free derivation of the sentence in question.

If your sentence is indeed logically valid, this method lets you know for sure. Unfortunately, if your sentence is not logically valid, you will never become aware of this fact! For, you will never be sure that, some day, a derivation of the required form will turn up.

A similar observation holds for the notion of being a logical consequence of a given concrete (more precisely: recursively enumerable) set of axioms.

An obvious question now is, whether this partial decidability result can be complemented by a method that allows you to ascertain non-logical validity in the same sense. If that would be possible, you could use computers to search for proofs (at least, in principle). However: the *Theorem of Church* (1936) states that this is impossible.

Logical validity for propositional formulas (that do not contain quantifiers) *is* decidable: for this, you only need to construct the *truth table* of the formula and inspect its last column. (Searching proof trees is not necessary here.) However, when your formula is composed using many atoms, this truth-table method may not be *feasible*: the truth table of an n-atom formula has 2^n rows, hence the time necessary to construct it is an exponential function of the length of the formula. Algorithmic methods are considered feasible only if they produce the required answer within an amount of time that is a *polynomial* function of the length of the input (i.e., the number of steps in the algorithmic computation is bound by a certain fixed power of the length of the input). This requirement defines the class P of *polynomial* problems. Satisfiability of propositional formulas is an example of a so-called NP problem. Cf. Theorem 3.34 and the explanation preceding it. The single most important open question for over 25 years in complexity theory (P = NP?) is whether NP problems like propositional satisfiability admit of a polynomial algorithmic solution. Propositional satisfiability happens to be NP-*complete*, which means that a polynomial solution for this special case can be transformed into a polynomial solution for any other of the many problems in NP. The informed guess is that propositional satisfiability is not polynomially solvable, but a proof of this is still lacking.

A.13 Computable sequence. A set of formulas is *computable* if there is an algorithmic method with which you can generate its elements one by one. An enumeration ψ_0, ψ_1, ψ_2, ... that is algorithmically produced is called a *computable* enumeration.

This notion of computability coincides with positive decidability. Indeed, to positively decide membership in a computable set, you just have to generate it and look whether the the object in question turns up. Conversely, if a set (of formulas) is positively decidable, you can obtain a computable enumeration of its elements (at least, if there are any) by the following method. Suppose that $\varphi_0, \varphi_1, \varphi_2, \ldots$ is an enumeration of *all* formulas. During one minute (say), let the computer figure out whether φ_0 is in the set. If the answer is positive, let φ_0 be the first element of the enumeration to be constructed. Otherwise, give the computer *two* minutes to figure out whether φ_0 is in the set, and the same amount of time to look into this question with respect to φ_1. As soon as you get a positive answer, you have a next element of the enumeration. Next, give the computer *three* minutes to see whether φ_0 is in the set (if this is still undecided), three minutes to see whether φ_1 is in it (likewise) and three minutes to work on this question with respect to φ_2. Going on this way, eventually all members of the set will be found and enumerated.

By the above discussion, the following is immediate.

A.14 Lemma. *If* Σ *is computable, then so is* $\{\varphi \mid \Sigma \models \varphi\}$.

Exercises

177 Show that every set of sentences with a model is consistent.
Hint. Apply the Soundness Theorem.

178 Derive the Consistency Theorem from the Completeness Theorem.

179 Derive the Completeness Theorem from the Consistency Theorem.
Hint. Assume that $\Gamma \not\vdash \varphi$. Use the rule RAA to show that $\Gamma \cup \{\neg\varphi\}$ is consistent.

180 Prove the Compactness Theorem using Corollary A.11 (which combines Consistency and Soundness Theorems).

181 Cf. the discussion in A.12. Describe a *feasible* decision method for satisfiability of propositional formulas in disjunctive normal form. (So, if $P \neq NP$, then calculating disjunctive normal forms cannot be feasible.)

A.3.2 Consistency Theorem

This section contains a proof of the Consistency Theorem A.9. The Completeness Theorem A.10 follows by Exercise 179. The proof is obtained from the one for the Compactness Theorem by replacing 'finitely satisfiable' by 'consistent'. In some places, this modification requires some rewriting.

Proof of the Consistency Theorem. Suppose that Γ is a consistent set of sentences. By Lemma A.16 there is a *maximally consistent Henkin set* $\Sigma \supset \Gamma$. By Lemma A.17, Σ has a model, and therefore Γ has a model as well. $\dashv$

A.15 Maximally Consistent, Henkin. A consistent set of sentences (in a given vocabulary) is *maximally consistent* if it is not a proper subset of a (different) consistent set of sentences in the *same* vocabulary.

Recall that a set of sentences Σ is *Henkin* if for every existentially quantified sentence $\exists x\varphi(x) \in \Sigma$ there is a constant symbol $\mathbf{c}$ such that $\varphi(\mathbf{c}) \in \Sigma$.

Note that the notion of maximal consistency is *relative to a vocabulary*, just like the notion of maximal finite satisfiability. Otherwise, no maximally consistent set would exist: if Σ is consistent and $\mathbf{c}$ a constant symbol not in a sentence from Σ, then Σ is a proper part of the set $\Sigma \cup \{\mathbf{c} = \mathbf{c}\}$ which is consistent as well.

A.16 Lemma. *Every consistent set of sentences has a maximally consistent superset that is Henkin (be it in an extended vocabulary).*

A.17 Lemma. *Every maximally consistent set of sentences that is Henkin has a model.*

The proofs of these lemmas require a number of other ones.

A.18 Lemma. *Every consistent set of sentences has a maximally consistent superset.*

Proof. Cf. the proof of Lemma 4.2. Apply Zorn's Lemma to the collection of consistent supersets, ordered by inclusion. $\dashv$

Compare the next Lemma to Lemma 4.4.

A.19 Lemma. *Every consistent set Σ of L-sentences can be extended to a consistent set Σ' of sentences in a vocabulary L' simply extending L such that*

$$\textit{if } \exists x\varphi(x) \in \Sigma, \textit{ then for some individual constant } \mathbf{c} \in L'\text{: } \varphi(\mathbf{c}) \in \Sigma'.$$

Proof. Add a new constant $\mathbf{c}_\varphi$ (a *witness*) for each existential L-sentence $\exists x\varphi(x)$ and define $\Sigma' := \Sigma \cup \{\varphi(\mathbf{c}_\varphi) \mid \exists x\varphi(x) \in \Sigma\}$. It must be shown that Σ' is consistent. This is slightly harder than the corresponding detail in the proof of Compactness. Again, a lemma is needed.

A.20 Lemma. *Suppose that Γ is a set of sentences, that $\varphi = \varphi(x)$, that $\mathbf{c}$ does not occur in φ or in a sentence of Γ, and that $\Gamma, \varphi(\mathbf{c}) \vdash \bot$. Then $\Gamma, \exists x\varphi(x) \vdash \bot$.*

Proof. Assume that $\mathcal{D}$ is a derivation of $\bot$ from Γ en $\varphi(\mathbf{c})$. Choose a variable z that does not occur in φ or in a formula in $\mathcal{D}$. In every formula of $\mathcal{D}$, replace $\mathbf{c}$ by z. The result is a derivation $\mathcal{D}^-$ of $\bot$ from Γ and $\varphi(z)$. (To see this, use induction on $\mathcal{D}$.)

Extend $\mathcal{D}^-$ as follows:

$$\dfrac{\dfrac{\dfrac{\dfrac{\dfrac{\begin{matrix}\Gamma \quad [\varphi(z)]^1 \\ (\textit{the derivation } \mathcal{D}^-) \\ \bot\end{matrix}}{\neg\varphi(z)}\text{D-1}}{\forall z\neg\varphi(z)}\text{G}}{\neg\varphi(x)}\text{I}}{\forall x\neg\varphi(x)}\text{G} \qquad \exists x\varphi(x)}{\bot}\text{MP}$$

This is a derivation of $\bot$ from Γ and $\exists x\varphi(x)$.

For the remainder of the proof, see Exercise 183. $\dashv$

The proof of Lemma A.19 is resumed.
A derivation of $\bot$ from Σ^+ can only use finitely many sentences as hypotheses. Therefore, assume that $\Sigma, \varphi_1(\mathbf{c}_1), \ldots, \varphi_m(\mathbf{c}_m) \vdash \bot$. Application of Lemma A.20 to $\Gamma := \Sigma \cup \{\varphi_1(\mathbf{c}_1), \ldots, \varphi_{m-1}(\mathbf{c}_{m-1})\}$ and $\varphi := \varphi_m$ shows that $\Sigma, \varphi_1(\mathbf{c}_1), \ldots, \varphi_{m-1}(\mathbf{c}_{m-1}), \exists x_m \varphi_m(x_m) \vdash \bot$. However, $\exists x_m \varphi_m(x_m) \in \Sigma$. Therefore, $\Sigma, \varphi_1(\mathbf{c}_1), \ldots, \varphi_{m-1}(\mathbf{c}_{m-1}) \vdash \bot$.

In the same way, $\varphi_1(\mathbf{c}_1), \ldots, \varphi_{m-1}(\mathbf{c}_{m-1})$ are eliminated one by one. Eventually, you find that $\Sigma \vdash \bot$. But this contradicts the consistency of Σ. $\dashv$

Proof of Lemma A.16. Alternate Lemmas A.18 and A.19 infinitely often. Cf. Exercise 182. $\dashv$

Compare the next result to Lemma 4.3.

A.21 Lemma. *Suppose that Σ is maximally consistent and Henkin. Then:*

1. $\Sigma \vdash \varphi \Rightarrow \varphi \in \Sigma$,
2. $\bot \notin \Sigma$,
3. $\neg\varphi \in \Sigma \Leftrightarrow \varphi \notin \Sigma$,
4. $\varphi \wedge \psi \in \Sigma \Leftrightarrow \varphi, \psi \in \Sigma$,
5. $\varphi \rightarrow \psi \in \Sigma \Leftrightarrow \varphi \notin \Sigma$ *or* $\psi \in \Sigma$,
6. $\exists x \varphi(x) \in \Sigma \Leftrightarrow \exists \mathbf{c}\, (\varphi(\mathbf{c}) \in \Sigma)$,
7. $\forall x \varphi(x) \in \Sigma \Leftrightarrow \forall \mathbf{c}\, (\varphi(\mathbf{c}) \in \Sigma)$.

Proof. 1. If $\Sigma \vdash \varphi$, then $\Sigma \cup \{\varphi\}$ is consistent.
3. If $\varphi \notin \Sigma$, then, by maximality, we have that $\Sigma, \varphi \vdash \bot$ and hence (Deduction rule) that $\Sigma \vdash \neg\varphi$.

See Exercise 184 for the remaining parts. $\dashv$

Proof of Lemma A.17. Compare the proof of Lemma 4.6.
Suppose that Σ is maximally consistent and Henkin. Define the relation $\sim$ on the set of variable-free terms by

$$s \sim t \equiv s = t \in \Sigma.$$

The universe A of the model $\mathcal{A}$ for Σ to be constructed is the set of equivalence classes $|t| := \{s \mid s \sim t\}$.

Over this universe we have to define constants, functions and relations corresponding to the constant, function and relation symbols of the vocabulary of Σ.

• *Constants.*
The interpretation $\mathbf{c}^{\mathcal{A}}$ of a constant symbol $\mathbf{c}$ is the corresponding equivalence class $|\mathbf{c}|$.

• *Functions.*
Suppose that $\mathbf{f}$ is an n-ary function symbol. Its interpretation in $\mathcal{A}$ is defined by $\mathbf{f}^{\mathcal{A}}(|t_1|, \ldots, |t_n|) := |\mathbf{f}(t_1, \ldots, t_n)|$.

It must now be verified that the (value of) the right-hand side of this definition of $\mathbf{f}^{\mathcal{A}}$ does not depend on the representatives chosen. Suppose that $|s_i| = |t_i|$ $(i = 1, \ldots, n)$. That is, $s_i \sim t_i$, i.e., $s_i = t_i \in \Sigma$ $(i = 1, \ldots, n)$. Then by Lemma A.21 we have that $\mathbf{f}(s_1, \ldots, s_n) = \mathbf{f}(t_1, \ldots, t_n) \in \Sigma$ (using an appropriate identity axiom), i.e., $\mathbf{f}(s_1, \ldots, s_n) \sim \mathbf{f}(t_1, \ldots, t_n)$, hence $|\mathbf{f}(s_1, \ldots, s_n)| = |\mathbf{f}(t_1, \ldots, t_n)|$.

Although the model $\mathcal{A}$ is not yet completely defined, you can already evaluate variable-free terms in it.

A.22 Lemma. *If t is a variable-free term, then $t^{\mathcal{A}} = |t|$.*

Proof. See Claim 1 of the proof of Lemma 4.6. ⊣

• *Relations.*
Suppose that $\mathbf{r}$ is an n-ary relation symbol. Its interpretation in $\mathcal{A}$ is defined by $\mathbf{r}^{\mathcal{A}}(|t_1|, \ldots, |t_n|) :\equiv \mathbf{r}(t_1, \ldots, t_n) \in \Sigma$. Again it must be verified that (the truth of) the right-hand side of this definition does not depend on the representatives chosen.

Finally, the proof is completed by thc

Claim. For every sentence φ we have that $\mathcal{A} \models \varphi$ iff $\varphi \in \Sigma$.

Proof. Almost identical to the one of Claim 2 in the proof of Lemma 4.6. For the remaining details, cf. Exercise 185. ⊣

Exercises

182 Fill in the details of the proof of Lemma A.16. Why is the set constructed consistent? Why is it *maximally* consistent? Why is it Henkin?
Hint. For the last two questions: a sentence of this set already belongs to the vocabulary of a certain stage in the construction.

183 In the proof of Lemma A.20 you need, at the instantiation-step, that x is substitutable for z in $\varphi(z)$. Why is that true? Why has $\mathbf{c}$ not been replaced by x *immediately* in $\mathcal{D}$ to obtain $\mathcal{D}^-$?
Hint for the last question. Problems both with instantiation and generalisation steps in $\mathcal{D}$.

184 Complete the proof of Lemma A.21.
Hint. Part 7 requires that $\neg\forall x\varphi \vdash \neg\forall x\neg\neg\varphi$.

185 Complete the proof of Lemma A.17. Verify that the definition of relations is independent of the representatives. Complete the proof of the *Claim.*
Warning. The logical constants here are $\bot$, $\rightarrow$, $\wedge$ and $\forall$.

186 In the proof of the Consistency Theorem it looks as if only *variable-free* identity axioms are being used. Why do you need identity axioms that contain variables?

187 ♣ Consider $\neg$ as a primitive, non-defined connective. Construct rules for $\neg$ that are sound and complete.

188 ♣ Consider $\vee$ as a primitive, non-defined connective. Construct rules for $\vee$ that are sound and complete.

189 In mathematical arguments the following pattern is often used. On the basis of a premise $\exists x\varphi(x)$ one chooses an object $\mathbf{c}$ for which $\varphi(\mathbf{c})$ holds. The rest of the argument uses this object; nevertheless, the final conclusion is supposed to follow on the basis of $\exists x\varphi(x)$ alone. Justify this pattern.

I.e., suppose that $\mathbf{c}$ does not occur in ψ or in a formula of Γ and that $\Gamma, \varphi(\mathbf{c}) \vdash \psi$. Show that $\Gamma, \exists x\varphi(x) \vdash \psi$.

This generalizes Lemma A.20. Avoid using the Completeness Theorem. *Hint.* Use Example A.4.3a.

190 In mathematical arguments one sometimes finds the following pattern. On the basis of a premise $\forall x\exists y\varphi(x,y)$ one postulates the existence of a function $\mathbf{f}$ such that $\forall x\varphi(x, \mathbf{f}(x))$ holds. The argument exploits this function. Nevertheless, the final conclusion is supposed to hold on the basis of the original premise. Justify this pattern.

I.e., suppose that $\mathbf{f}$ does not occur in ψ or in a formula of Γ, that $\mathbf{f}(x)$ (or x) is substitutable for y in $\varphi(x,y)$, and that $\Gamma, \forall x\varphi(x, \mathbf{f}(x)) \vdash \psi$. Show that $\Gamma, \forall x\exists y\varphi(x,y) \vdash \psi$.
Hint. Use the Completeness Theorem. By Exercise 19, to any model $\mathcal{A}$ of $\forall x\exists y\varphi(x,y)$ a (*Skolem*) function $f : A \to A$ can be found such that $(\mathcal{A}, f) \models \forall x\varphi(x, \mathbf{f}(x))$ (where f is taken as interpretation of $\mathbf{f}$).

Bibliographic Remark

The Completeness Theorem (be it for a different deduction apparatus) is due to Gödel. The, now classical, method of proof used here is from Henkin 1949.

B

Set Theory

This appendix summarizes the set-theoretic preliminaries that you are supposed to be familiar with when reading this book. But do not let it frighten you: most of the previous chapters use only very little.

B.1 Axioms

The standard system of first-order set-theoretic axioms is ZF, Zermelo-Fraenkel set theory. It has two primitives: the notion of a set, and the relation *to be a member of*, denoted by $\in$.

Usually, it is not too important what these axioms are about. The *Extensionality Axiom* states that sets with the same elements are equal. Most other axioms express that certain well-determined sets exist. For instance, to every two objects a and b there is the set $\{a, b\}$ the elements of which are exactly a and b (*Pairing Axiom*); every set a has a *sumset* $\bigcup a = \bigcup_{x \in a} x$ (*Sumset Axiom*) and a *powerset* $\mathcal{P}(a) = \{b \mid b \subset a\}$ (*Powerset Axiom*). Furthermore, a well-defined part $\{x \in a \mid E(x)\}$ of a set a always constitutes a set (its elements are those of a that satisfy the condition E) (Zermelo's *Aussonderung Axiom*, weakening Cantor's Comprehension Principle that was found contradictory) and the image of a set under a well-defined operation always constitutes a set (the *Replacement Axiom*, the Fraenkel-Skolem addition to the Zermelo axioms).

Every now and again, *classes* or *collections* are mentioned, such as the class of all sets, the class of all ordinals, etc. These objects are always assumed to be given by a defining property that determines membership in the class. Often (as is the case for the two examples mentioned) these things are not sets (they are *proper*).

B.2 Notations

The symbols $\subset$, $\cup$, $\cap$ and $-$ are used for set inclusion, union, intersection and set difference. The *cartesian product* $A \times B$ of A and B is the set of all pairs (a, b) where $a \in A$ and $b \in B$. More generally, the product of the sets A_i ($i \in I$), notation: $\prod_{i \in I} A_i$, is $\{f : I \to \bigcup_{i \in I} \mid \forall i \in I(f(i) \in A_i)\}$. If for all i, $A_i = A$, the product is a *power* A^I, the set of all functions from I to A.

Familiar examples of sets are $\mathbb{N} = \{0, 1, 2, \ldots\}$, the set of natural numbers, $\mathbb{Z}$, the set of integers, $\mathbb{Q}$ the set of rationals, and $\mathbb{R}$, the set of reals.

Often, a natural number n is identified with the set of its predecessors $\{0, \ldots, n-1\}$. This has the convenient effects that a natural number n has exactly n elements, and that the ordering coincides with membership.

B.3 Orderings

A relation $\preceq$ on a set A is a *partial ordering* if it is

- *reflexive*, for all $a \in A$, $a \preceq a$,
- *antisymmetric*, for all $a, b \in A$, if $a \preceq b$ and $b \preceq a$, then $a = b$,
- *transitive*, for all $a, b, c \in A$, if $a \preceq b$ and $b \preceq c$, then $a \preceq c$.

Such a partial ordering is *linear* if, additionally, for all $a, b \in A$, $a \preceq b$ or $b \preceq a$.

Sometimes, instead of these reflexive orderings, their irreflexive partners are used. The defining properties for an irreflexive ordering $<$ of A are *irreflexivity* (for no $a \in A$, $a < a$) and transitivity.

Orderings $(A, \preceq)$ constitute examples of models, for which we have the notion of *isomorphism*. An equivalence class of orderings under isomorphism is called an *order type*.

The following order types arise frequently:

- **n**, the order type of all n-element linear orderings,
- ω, the order type of $\mathbb{N}$ under its usual ordering,
- similarly, ζ, η and λ denote the order types of $\mathbb{Z}$, $\mathbb{Q}$ and $\mathbb{R}$, respectively, under their usual ordering.

Notation. We make a habit of confusing an ordering with its type. Thus, e.g., $\omega = (\mathbb{N}, <)$ (or $\omega = (\mathbb{N}, \leq)$, whatever is convenient).

If $(A, \preceq)$ is an ordering, its upside down version $(A, \succeq)$ is an ordering as well. It is denoted by $(A, \preceq)^\star$.

If $(A, \preceq)$ and $(B, \leq)$ are orderings on disjoint sets A and B, you can form their ordered sum $(A, \preceq) + (B, \leq)$, which is $A \cup B$ ordered by the relation that is the union of $\preceq$, $\leq$, and $A \times B$ (the relation in which every $a \in A$ precedes every $b \in B$). Thus, $\zeta = \omega^\star + \omega$.

More generally, you can form arbitrary ordered sums as follows. $\sum_{i \in I} \alpha_i$ is the ordered sum of (possibly infinitely many) pairwise disjoint orderings (or copies of those) α_i ($i \in I$) in which the ordering is dictated by that of I. Thus, if $\alpha_i = (A_i, <_i)$ and $<$ is the ordering of I, then the ordering relation of $\sum_{i \in I} \alpha_i$ is $(\bigcup_i <_i) \cup \bigcup_{i<j}(A_i \times A_j)$.

A product $\alpha \cdot \beta$ is defined as a sum of pairwise disjoint copies of α over an index set that is ordered in type β.

B.4 Ordinals

A relation ϵ is *well-founded* on U if every non-empty subset X of U has an ϵ-minimal element; that is, an element $x \in X$ such that for no $y \in X$ we have that $y \epsilon x$. Equivalently, the following induction principle holds: if $X \subset U$ is such that $\forall u \in U (\forall x \epsilon u (x \in X) \Rightarrow u \in X)$, then $X = U$. Equivalently still, ϵ is well-founded on U if there are no infinite sequences $\cdots \epsilon u_2 \epsilon u_1 \epsilon u_0$.

A *well-ordering* is a linear ordering that is well-founded. Examples of well-orderings are finite linear orderings and ω. The orderings ζ, η and λ are not well-orderings.

Suppose that $<$ well-orders the set A. If A is non-empty, it must have a least element, say 0. If $A \neq \{0\}$, then $A - \{0\}$ has a least element, say 1. Etc.; if A is infinite, it must have elements $0, 1, 2, 3, \ldots$ in that order. Possibly, $A \neq \{0, 1, 2, \ldots\}$, in which case there is a least element ω in $A - \{0, 1, 2, \ldots\}$. Going on with a self-explaining notation, we may find the following sequence of elements in A:

$$0,\ 1,\ 2,\ 3,\ \ldots,\ \omega,\ \omega+1,\ \omega+2,\ \ldots,\ \omega+\omega=\omega\cdot 2,\ \omega\cdot 2+1,\ \ldots,\ \omega\cdot 3,$$
$$\ldots,\ \omega\cdot\omega=\omega^2,\ \ldots\omega^3,\ \ldots,\ \omega^4,\ \ldots,\ \omega^\omega,\ \ldots,\ \omega^{\omega^\omega},\ \ldots$$

An *ordinal* is the type of a well-ordering. The notations used in the above sequence are generally used to denote ordinals.

Ordinals of finite orderings are usually identified with the corresponding natural numbers. As is the case with the natural numbers, it is convenient to identify an ordinal with the set of its predecessors. (In fact, this trick — due to von Neumann — provides a clever way to defining the notion of an ordinal in ZF set theory.)

Every non-zero ordinal is either a *successor* (of an immediate predecessor) or a *limit.* The first limit is ω, a few next ones are $\omega \cdot 2$, $\omega \cdot 3, \ldots, \omega^2$.

Every ordinal α below ω^ω can be uniquely written in the *Cantor Normal Form*

$$\alpha = \omega^{n_1} \cdot m_1 + \cdots + \omega^{n_k} \cdot m_k,$$

where the n_i and m_i are natural numbers and $n_1 > \cdots > n_k$.

A class of ordinals is *closed* if it contains every limit (i.e., union) $\bigcup_{\xi<\gamma} \alpha_\xi$ of its elements α_ξ, it is *unbounded* if it contains arbitrarily large ordinals. A class of ordinals is *closed below* (or *in*) an ordinal α if it contains every union of its elements that is $< \alpha$ and it is *unbounded below* α if it contains arbitrarily large ordinals $< \alpha$.

B.5 Cardinals

Sets are *equinumerous* if there is a one-to-one correspondence between their elements. A *cardinal* is an equivalence class modulo this relation. The cardinal of a set A is denoted by $|A|$.

The cardinal of an n-element set is usually identified with the corresponding natural number (and ordinal). An *initial number* is an infinite ordinal that is not equinumerous with a predecessor. The initials in their natural order are $\omega_0 = \omega, \omega_1, \omega_2, \ldots, \omega_\omega, \ldots$ The first initial ω is countable, the next one, ω_1, uncountable.

All ordinals in the listing displayed in Section B.4 are countable. The first uncountable ordinal ω_1 is (very much) bigger than

$$\omega \cdot 2, \omega^2, \omega^\omega, \omega^{\omega^\omega}, \ldots.$$

The cardinals of these initials (and, more generally, of infinite well-ordered sets) are *alephs*. The first few alephs are $\aleph_0$, $\aleph_1$, $\aleph_2$, ..., $\aleph_\omega$, ...

Cardinals can be summed, multiplied and raised to a power. Assuming the Axiom of Choice (see Section B.6), all infinite cardinals are alephs and the finite versions of these operations are essentially trivial: if κ and λ are infinite cardinals, then $\kappa + \lambda = \kappa \cdot \lambda = max(\kappa, \lambda)$.

A cardinal — and the corresponding initial ordinal — is *regular* if it cannot be written as a sum of smaller cardinals using an index set of smaller power. All alephs/initials with a non-limit index are regular. A regular aleph $\aleph_\gamma$ with a limit index γ is called *inaccessible*; it is *strongly inaccessible* if, additionally, it exceeds all powers 2^κ for $\kappa < \aleph_\gamma$.

B.6 Axiom of Choice

A function on the set of non-empty sets A is a *choice function for A* if it assigns to every set in A an element in that set. The *Axiom of Choice* states that every set of non-empty sets has a choice function.

There are many well-known equivalents of this principle. One that is usually very convenient is the

B.1 Lemma of Zorn. *Suppose that the non-empty set A is partially ordered by $\leq$ in such a way that every chain has an upper bound. Then A has a $\leq$-maximal element.* $\dashv$

Here, $B \subset A$ is a *chain* if it is linearly ordered by $\leq$; b is an *upper bound* of B if for all $x \in B$, $x \leq b$; and $a \in A$ is *maximal* if for no $x \in A$, $a < x$.

B.7 Inductive Definitions

Let A be a set. A function $\Gamma : \mathcal{P}(A) \to \mathcal{P}(A)$, mapping subsets of A to subsets of A, is called an *operator over A*. Such an operator is *monotone* if $X \subset Y \subset A \Rightarrow \Gamma(X) \subset \Gamma(Y)$.

B.2 Fixed Points. Let $\Gamma : \mathcal{P}(A) \to \mathcal{P}(A)$ be an operator. A subset $K \subset A$ of A is called

1. Γ-*closed* or *pre-fixed point* of Γ if $\Gamma(K) \subset K$,
2. Γ-*supported* or *post-fixed point* of Γ if $K \subset \Gamma(K)$,
3. *fixed point* of Γ if $\Gamma(K) = K$,
4. Γ-*inductive* if it is included in every Γ-closed set:
$$\Gamma(X) \subset X \Rightarrow K \subset X,$$
5. Γ-*co-inductive* if it contains every Γ-supported set:
$$X \subset \Gamma(X) \Rightarrow X \subset K.$$

An example of a monotone operator over $\mathbb{N}$ is the one that sends a set X to $\{0\} \cup \{n+1 \mid n \in X\}$. $\mathbb{N}$ is the only pre-fixed point, but there are plenty of post-fixed points. Mathematical Induction says precisely that $\mathbb{N}$ is inductive.

It is usually the least fixed point that is of interest. For instance, the sets of terms and formulas of a given vocabulary are least fixed points of suitable monotone operators over the set of expressions; another example occurs in the proof of Lemma 2.12. The least fixed point of an operator is said to be *inductively defined* by it. The notion of Ehrenfeucht game provides a natural operator Γ where the *greatest* fixed point is the interesting object: see the definition of Γ immediately after Lemma 3.20.

B.3 Lemma.

1. *There is at most one inductive pre-fixed point and at most one co-inductive post-fixed point.*
2. *Inductive pre-fixed points and co-inductive post-fixed points of a monotone operator are fixed points.*

Proof. The claims from the first sentence are obvious. Now assume that I is an inductive pre-fixed point of Γ. Then $\Gamma(I) \subset I$. By monotonicity, $\Gamma(\Gamma(I)) \subset \Gamma(I)$, that is: $\Gamma(I)$ is a pre-fixed point. By induction, $I \subset \Gamma(I)$. Therefore, I is a fixed point. The proof for co-inductive post-fixed points is similar. $\dashv$

By this lemma, inductive pre-fixed points are the same as inductive fixed points and as *least* fixed points. Similarly, co-inductive post-fixed points, co-inductive fixed points and greatest fixed points all amount to the same thing.

B.4 Proposition. *Every operator that is monotone has an inductive and a co-inductive fixed point.*

Proof. Let I be the set of $a \in A$ that are in *every* pre-fixed point. If $\Gamma(X) \subset X$, then $I \subset X$; thus (by monotonicity) $\Gamma(I) \subset \Gamma(X)$ and it follows that $\Gamma(I) \subset X$. Therefore, $\Gamma(I) \subset I$. Obviously, I is Γ-inductive.

The co-inductive post-fixed point consists of the set of $a \in A$ that are in some post-fixed point. $\dashv$

Both inductive fixed point and co-inductive fixed point can be approximated using the so-called *upward* or *least fixed point hierarchy*, respectively: the *downward* or *greatest fixed point hierarchy*. The upward hierarchy consists of the sequence of sets $\Gamma{\uparrow}\alpha$ (α an ordinal) recursively defined as follows.

$$\begin{aligned} \Gamma{\uparrow}0 &= \emptyset \\ \Gamma{\uparrow}(\alpha+1) &= \Gamma(\Gamma{\uparrow}\alpha) \\ \Gamma{\uparrow}\gamma &= \bigcup_{\xi<\gamma} \Gamma{\uparrow}\xi, \text{ if } \gamma \text{ is a limit ordinal.} \end{aligned}$$

The downward hierarchy consists of the sets $\Gamma{\downarrow}\alpha$ recursively defined in a similar way, but now starting from A instead of $\emptyset$ and taking intersections at limits.

B.5 Proposition. *Let Γ be monotone. $\Gamma{\uparrow} := \bigcup_\alpha \Gamma{\uparrow}\alpha$ is the least fixed point of Γ and $\Gamma{\downarrow} := \bigcap_\alpha \Gamma{\downarrow}\alpha$ is its greatest fixed point.*

Proof. Let I be the least fixed point. Then $\Gamma{\uparrow}0 \subset \cdots \subset \Gamma{\uparrow}\alpha \subset \cdots \subset I$, and so $\Gamma{\uparrow} \subset I$. Obviously, there is an ordinal α of power at most $|A|$ for which $\Gamma{\uparrow}\alpha = \Gamma{\uparrow}(\alpha+1)$. The least α for which this holds is the (upward) *closure ordinal* of Γ and we have that $\Gamma{\uparrow} = \Gamma{\uparrow}\alpha$ and, by Γ-induction, $I \subset \Gamma{\uparrow}$. $\dashv$

Example. Let ϵ be a relation on U. Define the monotone operator π over U by $\pi(X) := \{x \in U \mid \forall y \epsilon x (y \in X)\}$. The least fixed point of π is the *well-founded part* of U, that is: the largest initial part of U on which ϵ is well-founded.

In ZF set theory it is also possible to consider operations over proper classes, that map subclasses to subclasses.

An important example of such an operation is the *powerclass* operation that maps a class to the class of its sub*sets*. The stages of the least fixed point hierarchy of this operator are usually called *partial universes* and denoted by V_α. This *cumulative hierarchy* has no closure ordinal: every $V_{\alpha+1}$ has elements (such as the previous stage V_α and the ordinal α) that are not yet in V_α.

The *constructible hierarchy* $\{L_\alpha\}$ modifies the previous one in that a next stage $L_{\alpha+1}$ does not consist of *all* subsets of the previous L_α, but only of those that are parametrically first-order definable in the model $(L_\alpha, \in)$.

The least fixed point of the powerclass operator is the *well-founded part* of the universe. The *Regularity Axiom* of ZF expresses that this coincides with the universe itself.

If λ is an initial number of strongly inaccessible cardinality, it follows by induction that for $\alpha < \lambda$, $|V_\alpha| < |V_\lambda|$ and, hence, that $|V_\lambda| = |\lambda|$.

B.8 Ramsey's Theorem

B.6 Pigeon-hole Principle. *If $f : A \to I$, where A is infinite and I is finite, then infinitely many $a \in A$ will be mapped to the same $i \in I$.*

Proof. Suppose that every set $A_i := \{a \in A \mid f(a) = i\}$ were finite. Then $A = \bigcup_i A_i$ is a finite union of finite sets. However, a finite union of finite sets must be finite. (Induction on the number of elements of the index set.) $\dashv$

The following classical result about trees is an easy application of the Pigeon-hole Principle. A *tree* is a partial ordering $(T, <)$ with a least element r (the *root* of the tree) such that every subset of the form $\{x \in T \mid x < t\}$ $(t \in T)$ is finite and linearly ordered by $<$. Such a tree *splits finitely* if every element has at most finitely many immediate successors. A *branch* is a maximal linearly ordered subset $t_0 = r < t_1 < t_2 < \cdots$.

B.7 König's Lemma. *Every finitely splitting infinite tree has an infinite branch.*

Proof. The required branch $t_0 = r < t_1 < t_2 < \cdots$ is found by picking (for $n = 0, 1, 2, \ldots$) an immediate successor t_{n+1} of t_n with the property that $\{t \in T \mid t_{n+1} \leq t\}$ is infinite. Since the tree is finitely splitting and infinite, such an immediate successor exists by the pigeon-hole principle. $\dashv$

The next application of the Pigeon-hole Principle is more complicated.

B.8 Ramsey's Theorem. *If f maps the n-element subsets of an infinite set A to the elements of a finite set, then there exists an infinite subset of A the two-element subsets of which have the same f-image.*

Proof. Induction with respect to n. The case for $n = 1$ coincides with the Pigeon-hole Principle. Here follows the case for $n = 2$.

The proof starts with the construction of two infinite sequences: a sequence $a_0, a_1, a_2, \ldots \in A$, and a decreasing sequence $A_0 \supset A_1 \supset A_2 \supset \cdots$ of subsets of A. (The second sequence has a role in the construction of the first one only and will be disregarded afterwards.)
To begin with, take $A_0 := A$, and $a_0 \in A_0$ arbitrarily.

Consider the map on $A_0 - \{a_0\}$ defined by $a \longmapsto f(\{a_0, a\})$. By the Pigeon-hole Principle there exists an infinite subset $A_1 \subset A_0 - \{a_0\}$ of elements all having the same image. Let a_1 be an arbitrary element of A_1.

Next, define a map on $A_1 - \{a_1\}$ by $a \longmapsto f(\{a_1, a\})$. Again, there is an infinite subset $A_2 \subset A_1 - \{a_1\}$ of elements all with the same image.

Continue in this fashion. Note that eventually we have, for the sequence $a_0, a_1, a_2, \ldots$ that, if $i < j < k$, then (since $a_j, a_k \in A_{i+1}$) $f(\{a_i, a_j\}) = f(\{a_i, a_k\})$. I.e.: the f-value of a two-element subset of $\{a_0, a_1, a_2, \ldots\}$ only depends on the element that has the *least* index.

Define a last map on $\{a_i \mid i \in \mathbb{N}\}$ by $a_i \longmapsto f(\{a_i, a_{i+1}\})$.

By a final application of the Pigeon-hole Principle, obtain an infinite set $H \subset A$ of elements with the same image. This set has the desired property: indeed, if $a_i, a_j, a_k, a_l \in H$, $i < j$ en $k < l$, then we have that $f(\{a_i, a_j\}) = f(\{a_i, a_{i+1}\}) = f(\{a_k, a_{k+1}\}) = f(\{a_k, a_l\})$. $\dashv$

B.9 Games

You are undoubtedly familiar with several examples of two-person games. The following definitions involving the *tree of all positions* of a given game try to make the intuitive notion more precise.

A *finite two-person game* — where one can only win or lose and a draw is impossible — can be given as a tree-like structure on the set T of *positions* that occur in all possible plays of the game. There is an *initial position* $r \in T$. For every *non-final* position $t \in T$ it is determined whose turn it is to move and to which positions the player whose turn it is can move to. The notation $t \mapsto t'$ indicates that, in position t, the player whose turn it is can move to t'. In *final* positions the game is over, no move is possible, and one of the players is designated as the *winner*. A *play* of the game is a maximal sequence of positions $t_0 = r, t_1, t_2, \ldots$ that starts at r and is such that every transition $t_i \mapsto t_{i+1}$ is a legal move.

Games you play in daily life are *finite* in the sense that all plays have finite length. However, also an infinite game will be considered. This means that no final positions exist and every play of the game continues indefinitely. In this case it is the *plays* themselves that are partitioned into those in which the first player has won and those in which the second player has won.

B.9 Strategies. A *strategy* for one of the players is a prescription how to play. A strategy is *winning* if, by following it, you win every play of the game, no matter your opponent's moves.

B.10 Determinacy of Finite Games. *For every finite game, exactly one of the players has a winning strategy.*

Proof. First, the two players cannot both have a winning strategy.

Next, let Win_X, the *winning set* for X, be the set of positions in which player X has a winning strategy (for the remaining part of the game). Proposition B.10 asserts that the root position r is in exactly one of these winning sets. To prove this, something stronger will be shown, namely,

Claim. The winning sets form a partition of T.

Proof. Note first that the sets Win_X can be inductively defined by the following rules.

(1) If X is designated as winner in the final position t, then $t \in Win_X$,

(2a) if it's X's turn to move in t and for some move $t \mapsto t'$, $t' \in Win_X$, then (since X can move to t' and win) $t \in Win_X$,

(2b) if it isn't X's turn to move in t and for all moves $t \mapsto t'$, $t' \in Win_X$, then (since his opponent cannot help moving to some $t' \in Win_X$) $t \in Win_X$.

By assumption, there are no infinite plays. Thus (see Section B.4), we can argue by induction along the (converse) relation $t \mapsto t'$.
(i) t is a final position.
This case is trivial, since by assumption, the final positions are partitioned in those won by the first and those won by the second player.
(ii) t is not final.
By induction hypothesis, if $t \mapsto t'$ is a move, then t' is in exactly one of the winning sets. Assume that it's X's turn to move in t.
(iia) For some move $t \mapsto t'$, $t' \in Win_X$. Then by (2a), $t \in Win_X$.
(iib) For no move $t \mapsto t'$, $t' \in W_X$. By induction hypothesis, for all such t', $t' \in W_Y$, where Y is the other player. Then by (2b), $t \in Win_Y$. $\dashv$

The argument of this proof does not take into account that all plays have the same length. You can apply it (as did Zermelo) to the game of chess and conclude that either white or black "has" a strategy with which he cannot loose. The version of this result for infinite games still holds if you assume the game to be *open*. A set of plays is *open* if every play in it has a finite initial such that all plays that prolong this initial are also in the set. The game is *open* if there is a player whose set of winning plays is open.

191 Exercise. Show that open games are determined.
Hint. Assume (1) that it is the first player whose set of wins is open and suppose (2) that the second player does not have a winning strategy. Consider the strategy for the first player that consists in trying to avoid positions in which the second player has a winning strategy. Show: by assumption (2), the first player can indeed avoid those positions, and by assumption (1), doing so he must win.

Bibliographic Remarks

For an introduction to inductive definitions, see Aczel 1977. The least fixed point result goes back to Tarski 1955.

Ramsey's Theorem B.8 is from Ramsey 1928.

Determinacy of open games is due to Gale and Stewart 1953. The strongest determinacy result that is provable in ZF is due to Martin: it says that *Borel* games (where one of the players has a winning set that is Borel in the topology described) are determined. The *Axiom of Determinacy* states that every game (with a tree of finite degree) is determined. This curious assumption contradicts the Axiom of Choice and implies that all sets of reals are Lebesgue measurable. (Weakenings of this axiom stronger than Borel determinacy can leave some of AC intact and are related to large cardinals.)

Bibliography

Aczel, P. 1977. An introduction to inductive definitions. In *Handbook of Mathematical Logic*, ed. J. Barwise. 739–782. North Holland.

Ajtai, M., and R. Fagin. 1990. Reachability is harder for directed than for undirected finite graphs. *Journal of Symbolic Logic* 55:113–150.

Barwise, J. 1975. *Admissible Sets and Structures*. Springer-Verlag.

Barwise, J. 1985. Model-theoretic logics: background and aims. In *Model-Theoretic Logics*, ed. J. Barwise and S. Feferman. 3–23. Springer-Verlag.

Barwise, K.J., and J. Schlipf. 1976. An introduction to recursively saturated and resplendent models. *Journal of Symbolic Logic* 41:531–536.

Beth, E.W. 1953. On Padoa's method in the theory of definition. *Indagationes Mathematicae* 15:330–339.

Blass, A., Y. Gurevich, and D. Kozen. 1985. A zero-one law for logic with a fixed point operator. *Information and Control* 67:70–90.

Chang, C.C., and H.J. Keisler. 1990. *Model Theory*. North Holland. 3rd edition.

Craig, W. 1957. Three uses of the Herbrand-Gentzen Theorem in relating model theory and proof theory. *Journal of Symbolic Logic* 22:269–285.

Doets, K. 1987. On n-equivalence of binary trees. *Notre Dame Journal of Formal Logic* 28:238–243.

Ebbinghaus, H.-D. (ed.). 1987. *Model Theory*. Ω-Bibliography of Mathematical Logic, Vol. III. Berlin: Springer.

Ebbinghaus, H.-D. and Flum, J. 1995. *Finite Model Theory*. Perspectives in Mathematical Logic. Berlin: Springer.

Ehrenfeucht, A. 1961. An application of games to the completeness problem for formalized theories. *Fundamenta Mathematicae* 49:129–141.

Fagin, R. 1974. Generalized first-order spectra and polynomial-time recognizable sets. In *Complexity of Computation, SIAM-AMS Proceedings, Vol. 7*, ed. R. M. Karp, 43–73.

Fagin, R. 1975. Monadic generalized spectra. *Zeitschrift für Mathematische Logik und Grundlagen der Mathematik* 21:89–96.

Fagin, R. 1976. Probabilities on finite models. *Journal of Symbolic Logic* 41:50–58.

Fagin, R. 1990. Finite-model theory — a personal perspective. In *Proc. 1990 International Conference on Database Theory*, ed. S. Abiteboul and P. Kanellakis, 3–24. Springer-Verlag Lecture Notes in Computer Science 470. To appear in *Theoretical Computer Science.*

Fagin, R., L.J. Stockmeyer, and M.Y. Vardi. 1992. A simple proof that connectivity separates existential and universal monadic second-order logics over finite structures. Research Report RJ 8647. IBM.

FMT 1995. *References for Finite Model Theory.* Anonymous FTP address `ftp.informatik.rwth-aachen.de`, directory `pub/FMT/bib`. Addresses on the World-Wide Web: `http://math.wisc.edu/~quesada/refs/index.html`, and `http://www.informatik.rwth-aachen.de/WWW-math/FMT.html`.

Fraïssé, R. 1954. Sur quelques classifications des systèmes de relations. *Publ. Sci. Univ. Alger.* Série. A 1:35–182.

Gaifman, H., and M.Y. Vardi. 1985. A simple proof that connectivity is not first-order. *Bulletin of the European Association for Theoretical Computer Science* 26:43–45.

Gale, D., and F.M. Stewart. 1953. Infinite games of perfect information. *Annals of Mathematical Studies* 28:245–266.

Gurevich, Y. 1992. Zero-one laws. *Bulletin of the European Association for Theoretical Computer Science* 46:90–106.

Henkin, L. 1949. The completeness of the first-order functional calculus. *Journal of Symbolic Logic* 14:159–166.

Hodges, W. 1993. *Model Theory.* Encyclopdia of Mathematics and its Applications, Vol. 42. Cambridge, UK: Cambridge University Press.

Immerman, N. 1987a. Expressibility as a complexity measure: results and directions. In *Second Structure in Complexity Conference*, 194–202.

Immerman, N. 1987b. Languages that capture complexity classes. *SIAM Journal of Computing* 16:760–778.

Immerman, N. 1989. Descriptive and computational complexity. In *Computational Complexity Theory, Proc. Symp. Applied Math., Vol. 38*, ed. J. Hartmanis. 75–91. American Mathematical Society.

Immerman, N., and D. Kozen. 1989. Definability with bounded number of bound variables. *Information and Computation* 83:121–139.

Karp, C. 1965. Finite quantifier equivalence. In *The Theory of Models*, ed. J. Addison, L. Henkin, and A. Tarski. 407–412. North Holland.

Keisler, H. J. 1971. *Model Theory for Infinitary Logic.* North Holland.

Lindström, P. 1969. On extensions of elementary logic. *Theoria* 35:1–11.

Lyndon, R.C. 1959a. An interpolation theorem in the predicate calculus. *Pacific Journal of Mathematics* 9:155–164.

Lyndon, R.C. 1959b. Properties preserved under homomorphism. *Pacific Journal of Mathematics* 9:143–154.

Montague, R.M., and R.L. Vaugt. 1959. Natural models of set theories. *Fundamenta Mathematicae* 47:219–242.

Ramsey, F.P. 1928. On a problem in formal logic. *Proceedings of the London Mathematical Society* 30:264–286.

Ressayre, J.P. 1977. Models with compactness properties relative to an admissible language. *Annals of Mathematical Logic* 11:31–55.

Rosenstein, J.G. 1982. *Linear Orderings*. New York: Academic Press.

Ryll-Nardzewski, C. 1959. On the categoricity in power $\aleph_0$. *Bull. Acad. Polon. Sci. Sér. Sci. Math. Astron. Phys.* 7:239–263.

Svenonius, L. 1959. $\aleph_0$-categoricity in first-order predicate calculus. *Theoria* 25:82–94.

Tarski, A. 1955. A lattice theoretical fixpoint theorem and its applications. *Pacific Journal of Mathematics* 5:285–309.

Tarski, A., and R.L. Vaught. 1957. Arithmetical extensions of relational systems. *Compositio Mathematicae* 13:81–102.

Name Index

Subject Index

Notation

Set Theory

Other

Studies in Logic, Language and Information

a series co-published by

The European Association for Logic, Language and Information

The *Studies in Logic, Language and Information* book series is the official book series of the European Association for Logic, Language and Information (FoLLI).

The scope of the book series is the logical and computational foundations of natural, formal, and programming languages, as well as the different forms of human and mechanized inference and information processing. It covers the logical, linguistic, psychological and information-theoretic parts of the cognitive sciences as well as mathematical tools for them. The emphasis is on the theoretical and interdisciplinary aspects of these areas.

The series aims at the rapid dissemination of research monographs, lecture notes and edited volumes at an affordable price.

Other titles in the series include:

Logic Colloquium '92

László Csirmaz, Dov M. Gabbay, and Maarten de Rijke, editors

This volume is a collection of papers based on presentations given at Logic Colloquium '92. The contributions focus on the interaction between formal systems in logic and algebra. It contains both an up-to-date introductory overview of the area, as well as more specialized case studies, and papers on closely related fields.

336 p. ISBN: 1-881526-97-6 (cloth) ; 1-881526-98-4 (paper)

Meaning and Partiality

Reinhard Muskens

Philosophers, logicians, linguists and other researchers in artificial intelligence will find the development of a theory of meaning in ordinary language in this volume useful. The theory set forth by Muskens is strictly formalised on the basis of formal logic and this volume contributes to the general discipline of Formal Semantics. Within this field it offers a synthesis between two leading paradigms, Montague Semantics and Situation Semantics.

152 p. ISBN: 1-881526-80-1 (cloth) ; 1-881526-79-8 (paper)

Logic and Visual Information

Eric M. Hammer

This book examines the logical foundations of visual information: information presented in the form of diagrams, graphs, charts, tables, and maps. The importance of visual information is clear from its frequent presence in everyday reasoning and communication, and also in computation.

136 p. ISBN: 1-881526-87-9 (cloth) ; 1-881526-99-2 (paper)

Partiality, Modality, and Nonmonotonicity

Patrick Doherty, editor

This edited volume of articles provides a state of the art description of research in logic based approaches to knowledge representation which combines approaches to reasoning with incomplete information that include partial, modal, and nonmonotonic logics. The collection contains two parts: foundations and case studies. The foundations section provides a general overview of partiality, multivalued logics, use of modal logic to model partiality and resource-limited inference, and an integration of partial and modal logics. The case studies section provides specific studies of issues raised in the foundations section.

312 p. ISBN: 1-57586-031-7 (cloth) ; 1-57586-030-9 (paper)

Another title of interest:

Vicious Circles

Jon Barwise and Larry Moss

The subject of non-wellfounded sets came to prominence with the 1988 publication of Peter Aczel's book on the subject. Since then, a number of researchers in widely differing fields have used non-wellfounded sets (also called "hypersets") in modeling many types of circular phenomena. The application areas range from knowledge representation and theoretical economics to the semantics of natural language and programming languages.

Vicious Circles offers an introduction to this fascinating and timely topic. Written as a book to learn from, theoretical points are always illustrated by examples from the applications and by exercises whose solutions are also presented. The text is suitable for use in a classroom, seminar, or for individual study.

In addition to presenting the basic material on hypersets and their applications, this volume thoroughly develops the mathematics behind solving systems of set equations, greatest fixed points, coinduction, and corecursion. Much of this material has not appeared before. The application chapters also contain new material on modal logic and new explorations of paradoxes from semantics and game theory.

396 p. ISBN: 1-57586-009-0 (cloth) ;
1-57586-008-2 (pbk.)